How to Do *Everything* with

Adobe Encore DVD

C000022195

How to Do
Everything
with

Adobe®
Encore™ DVD

Doug Sahlin

McGraw-Hill/Osborne

New York   Chicago   San Francisco   Lisbon
London   Madrid   Mexico City   Milan   New Delhi
San Juan   Seoul   Singapore   Sydney   Toronto

*The **McGraw·Hill** Companies*

McGraw-Hill/Osborne
2100 Powell Street, 10th Floor
Emeryville, California 94608
U.S.A.

To arrange bulk purchase discounts for sales promotions, premiums, or fund-raisers, please contact **McGraw-Hill**/Osborne at the above address. For information on translations or book distributors outside the U.S.A., please see the International Contact Information page immediately following the index of this book.

How to Do Everything with Adobe® Encore™ DVD

1234567890 CUS CUS 019876543

ISBN 0-07-223190-4

Publisher:	Brandon A. Nordin
Vice President &	
Associate Publisher	Scott Rogers
Acquisitions Editor	Megg Morin
Acquisitions Coordinator	Athena Honore
Project Editor	Jennifer Malnick
Technical Editor	Joe Bowden
Copy Editor	Andrea Boucher
Proofreader	Marian Selig
Indexer	Claire Splan
Composition	International Typesetting & Composition
Illustrators	International Typesetting & Composition
Series Design	Mickey Galicia
Cover Illustration	Tom Willis
Cover Series Design	Dodie Shoemaker

This book was composed with Corel VENTURA™ Publisher.

Dedicated to the creative spirit that lives within us all.

About the Author

Doug Sahlin is a graphic and web site designer living in Central Florida. He is the author of 12 books on computer graphics and web animation including *How To Do Everything with Macromedia Contribute* and *How To Do Everything with Adobe Acrobat 6.0*, which was recently ranked 22nd in Amazon's top 50 computer books. He has authored online Flash courses and presented on-location seminars.

Doug has also coauthored a book on digital video; he is an accomplished photographer, videographer, and video editor, skills that he uses to augment his clients' web sites, multimedia applications, and DVD presentations.

Contents at a Glance

Contents

Acknowledgments

An endeavor of this magnitude would not have been possible without a vast support team. Many thanks to all the fine folks at McGraw-Hill/Osborne, especially Acquisitions Editor Megg Morin for making this project possible. A bouquet of virtual tulips and a bowlful of chicken soup to Project Editor Jenny Malnick for keeping my inbox full and translating Yiddish terms of endearment. Special thanks to Acquisitions Coordinator Athena Honore for making sure the chapters and accompanying screenshots were forwarded to the right parties. Thanks to the talented and very blonde copyeditor Andrea Boucher for not having any brunette moments while polishing this text for public consumption.

Kudos to technical editor Joe Bowden, whose input and insightful comments helped keep this book on track. Thanks to the Adobe Encore DVD design team and Adobe Video Evangelist Daniel Brown. As always, thanks to the lovely and talented Margot Maley Hutchison for providing liaison between publisher and author. Many thanks to my friends and family for their support, with special thanks to Karen and Ted.

Introduction

Adobe Encore DVD is a Windows application used to author DVDs. If you do not own a Windows computer, quietly put this book back on the shelf, as the application will not work on a Macintosh machine. But if you do own a Windows machine, by all means read on.

Adobe Encore DVD 1.0 is a brand-new application. But the name Adobe ensures you it's a rock-solid, well-thought-out application with the necessary tools and menu commands to author DVDs. If you use the application to its fullest, you'll create more than just garden-variety DVDs; you'll create compelling, eye-catching DVDs with features such as motion menus, animated buttons, subtitle tracks, alternate audio tracks, and much more. However, this degree of sophistication doesn't come without a learning curve.

That's where this book comes into play. *How To Do Everything with Adobe Encore DVD* is designed to help you get up and running quickly while getting the most from the application. The book will help users of all levels understand digital video and DVD authoring. In addition to finding a full-course serving of Adobe Encore DVD, you'll also find useful snippets of information on related topics such as shooting digital video for use with a DVD project, working with sound for DVD projects, replicating DVDs, and much more.

The information is presented in a logical progression starting with some necessary information about digital video and digital video formats as well as information about the DVD format and television broadcast standards. This preliminary information is followed with the information you need to author a DVD project with Adobe Encore DVD.

Each chapter builds on the previous, in the same progression that you'll use the various facets of the application to author DVD projects. The first chapters in the book show you how to organize your project, import assets, and create timelines, and is followed by chapters that show you how to use the application and your creative talent to author state-of-the-art DVD discs. The information is presented with a minimum of technical jargon; where technical information is presented, it is condensed into easy-to-understand terms.

The Structure of This Book

This book is organized into the following parts:

Part I: Introducing Adobe Encore DVD

In the first part of this book, you'll find information on the DVD format, digital video, and television broadcast standards used in the Americas and the rest of the world. You'll also receive an introduction to Adobe Encore DVD as well as a glimpse at the steps you'll use to author a DVD disc. In addition, you'll find a chapter that delves into all facets of the Adobe Encore DVD workspace (Chapter 3) with additional information that shows you how to customize the workspace to suit your workflow.

Part II: Authoring a DVD

This part of the book gets into the nitty-gritty of authoring a DVD project. You'll learn how to create a new project and specify project settings for the part of the world in which your DVD will be viewed, and you'll learn how to work with and organize project assets. In addition, you'll find information on how to properly plan your DVD project to avoid any potential pitfalls when you're deep into a project. In this part you'll find an entire chapter (Chapter 6) devoted to creating timelines for your video and audio assets, as well as creating chapter points and poster frames for each part of your DVD disc.

Part III: Working with Menus and Buttons

When your DVD is loaded into a set-top DVD player, the only visible parts of your handiwork are the menus, buttons, and, of course, the video that is linked to each menu button. Even though you do devote a considerable amount of time to planning your project and working with assets, if your menus are not eye-catching, and they have buttons that don't function properly, your finished DVD will fall short of the mark. In this part of the book, you'll find everything you need to know about creating flawless menus that pique viewer interest. You'll also learn how to properly link buttons to video clips and other menus. If you want to create a DVD disc that stands out from the crowd, you'll find the chapter on motion menus (Chapter 10) essential.

Part IV: Advanced DVD Techniques

If you've explored the application prior to purchasing this book, you know that Adobe Encore DVD ships with an impressive array of menus and buttons. However, if you want to put your own stamp of originality on a project, you need to create a custom menu, a skill you'll learn in the chapter devoted to creating custom DVD menus in Adobe Photoshop (Chapter 11). If your DVD needs features such as subtitles and alternate audio tracks, you'll find a chapter devoted to each in this part of the book.

Part V: Creating the DVD Disc

The end result of any DVD project is creating a disc. In this part of the book you'll find out how to do this in Adobe Encore DVD. You'll learn how to burn a project to a DVD disc, create a DVD image, and create a DVD master. You'll also find a chapter devoted to preparing your project for output (Chapter 14), in which you'll find useful information to ensure your DVD projects are error free. Information on copy-protecting your DVD discs can also be found in this part of the book.

Appendix A: Create Your Own DVD: A Step-By-Step Tutorial

This appendix encapsulates each phase of creating a DVD project in a condensed format. It reinforces the material presented in each part of the book and serves as a quick reference guide on how to create a DVD from start to finish.

Appendix B: Adobe Encore DVD Keyboard Shortcuts

This appendix lists each Adobe Encore DVD keyboard shortcut you can use to streamline your workflow. If you prefer the convenience of keyboard shortcuts, this appendix is a must read.

Appendix C: Internet DVD and Video Resources

The Internet is a treasure trove of useful information on most any topic, and DVD is no different. However, finding *useful* web sites about a topic can often be a daunting task. This appendix separates the wheat from the chaff and lists some beneficial Internet resources for creating DVDs and working with digital video.

Conventions Used In This Book

Whenever you find instructions to use a menu command, you'll see the path to the command listed. For example, when you see instructions to import an asset, the path to the Adobe Encore DVD menu command is listed as follows: File | Import

As Asset. When new terms are introduced, they are *italicized*. You will find a parenthetical reference for new terms, or a brief explanation. Whenever information on how to perform a detailed task is presented, it is listed in step-by-step format.

Throughout the book you'll find useful Tips that show you ways to streamline your workflow and Notes that alert you to certain issues or pitfalls to avoid. You'll find insights and interesting tidbits about related Adobe Encore DVD subjects in the Did You Know sidebars. You'll also find a generous helping of How to... sidebars, jam-packed with information on how to create elements for your Adobe Encore DVD projects.

Conclusion

This book is designed to be a desktop reference for creating DVDs with Adobe Encore DVD. You can read the book from cover to cover, or use the index to locate information pertaining to a specific task. I hope you find the information useful in your goal to get the most from Adobe Encore DVD and create state-of-the-art DVDs for yourself, your clients, or your organization.

Part I

Introducing Adobe Encore DVD

Chapter 1

About DVD and Digital Video

How to...

- Work with digital video
- Work with digital audio
- Select the proper DVD media
- Choose the proper broadcast standard
- Understand DVD terms and concepts

The advent of powerful computers and sophisticated software has radically changed the manner in which video entertainment is produced and delivered. It is now possible to create a full-fledged video production on a computer. Videographers have powerful nonlinear video editing applications at their disposal that can be used to create anything from a 30-second video advertisement to a full-length motion picture. This proliferation of technology has resulted in a cornucopia of video entertainment in every genre imaginable from low-budget thrillers like *The Blair Witch Project* to sophisticated, high-end animated productions *Monsters, Inc.* and *Toy Story*. Technology has also changed the way video entertainment is delivered.

In today's frenetic, fast-paced world, people are working longer and harder than ever before. Video entertainment is often the perfect antidote for a stressful and hectic day. Many people prefer the luxury of watching their video entertainment in the comfort of their own homes rather than being shoehorned into a crowded movie theater. Modern technology and fierce competition have made it possible for many households to upgrade from outdated VHS technology to moderately priced DVD players. The video quality of DVD is far superior to that offered on VHS tapes. As an added bonus, the discs are less fragile than VHS tapes and less prone to damage. This switch to DVD has led to an increased demand for entertainment in the DVD format, as well as educational DVD content, which has in turn led to an increased demand for authors of DVD material—the very reason you purchased Adobe Encore DVD (and, for that matter, this book).

This chapter will familiarize you with the aspects of digital video, DVD media, and the DVD format. While it is possible to do menial video-editing tasks such as trimming video clips in Adobe Encore DVD, you have to prepare your video content in video-editing applications such as Adobe Premiere Pro or Sony Vegas prior to authoring a DVD project or adding titles, transitions, and other video effects. In this regard, it's important to know which video and audio formats are supported by Adobe Encore DVD and to gain a working knowledge of digital video and the DVD format, topics that will be discussed at length in this chapter.

About Digital Video Formats

Digital video can be displayed on a wide variety of devices: cellular phones, PDAs, computer CD-ROM drives, web sites, and DVD players. When videographers render digital video, they choose a format that is supported by the device on which the video will be played. Choosing the proper format implies knowledge of video formats, audio formats, compression codecs, and so on. A codec is an alogrithm that compresses a video file when it is rendered in a video editing application. The same codec is used on a playback device to decompress the video. Applications like Adobe Premiere Pro and Sony Vegas enable you to edit digital video and add special effects, titles, and so on. Both applications support the vast majority of video formats. Other applications like Discreet Cleaner and Sorenson Squeeze specialize in compressing digital video for specific playback devices in one or more of the popular digital video formats. By combining video-editing and video compression applications, video producers can create productions that span the wide variety of video and audio formats from Macintosh AIFF audio to Windows WMV video format. Adobe Encore DVD supports a wide variety of digital video and audio formats, which will be discussed at length in Chapter 2.

Understand the MPEG-2 Format

MPEG stands for *Motion Pictures Experts Group*. Digital video for DVD discs are formatted in the MPEG-2 format. Like all digital video, MPEG-2 is compressed for the device type on which the video will be displayed. The DVD format permits a maximum data rate of 10,000 kbps. Video can be encoded with either VBR (Variable Bit Rate) or CBR (Constant Bit Rate).

When video editors render an MPEG-2 movie with VBR encoding, they specify the minimum, maximum, and target bit rates. The video editing application determines the actual bit rate based on the complexity of the material in each frame rendered. Complex video information with multiple images, complex scene transitions, and millions of colors is encoded near the maximum specified bit rate, while video information such as static titles on a solid color background is rendered near the minimum bit rate. All other video information is rendered at a bit rate between the maximum and minimum. Choosing VBR encoding ensures the smallest possible file size for the rendered video.

When video editors select CBR encoding, they specify the data rate. The video editing application renders every frame of the movie at the specified data rate. Remember that lower data rates enable you to fit more video on a DVD disc; however, the video quality will suffer. The optimum data rate is a compromise between the amount of video you need to pack on the DVD and the desired image quality.

You can import DVD-compliant MPEG-2 video files into Adobe Encore DVD as project assets. When you import an MPEG-2 video into Encore, the file does not need to be transcoded because MPEG-2 is the native video format for DVD discs. When you build a DVD project, the MPEG-2 video is transformed into a VOB (Video Object) that is stored in the disc's Video_TS directory.

Understand the AVI Format

When you capture video in a nonlinear video editing application, it is often captured as an AVI *(Audio Video Interleaved)* file, which was developed by Microsoft. AVI stores video and audio information and is a derivative of the RIFF *(Resource Interchange File Format)* format. You can import DVD-compliant AVI files directly into Adobe Encore DVD. You will, however, have to transcode these files prior to building a DVD project. You can transcode a file directly in Adobe Encore DVD, a task that will be discussed in detail in Chapter 7. When you transcode a file in Adobe Encore DVD, the video is transcoded into the MPEG-2 format, and the audio is transcoded into your choice of Dolby Digital sound, MPEG sound, or PCM sound.

 Understand MPEG-1 Video

MPEG-1 is the predecessor of MPEG-2. MPEG-1 does not offer the excellent image quality and high data rate of MPEG-2. MPEG-1 image quality is close to the image quality of a commercial VHS tape. Currently, MPEG-1 is used to create VCD *(Video Compact Disc)* discs. VCD discs hold either 650MB or 700MB of video data. MPEG-1 videos for VCD discs are sized at 352×288 pixels (PAL broadcast standard) or 352×240 pixels (NTSC broadcast standard). When encoded into the VCD format and played back on a DVD player, the video is increased to the DVD size of 704×576 pixels (PAL broadcast standard) or 720×480 pixels (NTSC broadcast standard). Adobe Encore DVD does not support the MPEG-1 format, and therefore does not support the authoring of VCD discs.

If a client presents you with video in the MPEG-1 format, advise the client that the video cannot be used for a DVD project. Request that the client present you with the original video if available. You can then capture the video into an application such as Adobe Premiere Pro, and then export the video in a format that can be used by Adobe Encore DVD.

If you edit your video projects using a nonlinear video-editing application such as Adobe Premiere Pro or Sony Vegas, you can export your video project in the AVI format. For the optimal image quality, you should render the file as an uncompressed AVI file. However, when you work with uncompressed AVI files, the file sizes are huge and playback performance in Adobe Encore DVD may suffer. In this regard, when you are rendering a video file for use in Encore DVD, use a DVD compliant rendering template from your video editing application. The following illustration shows the dialog box from a video editing application with settings for rendering an uncompressed AVI file for use in a DVD project:

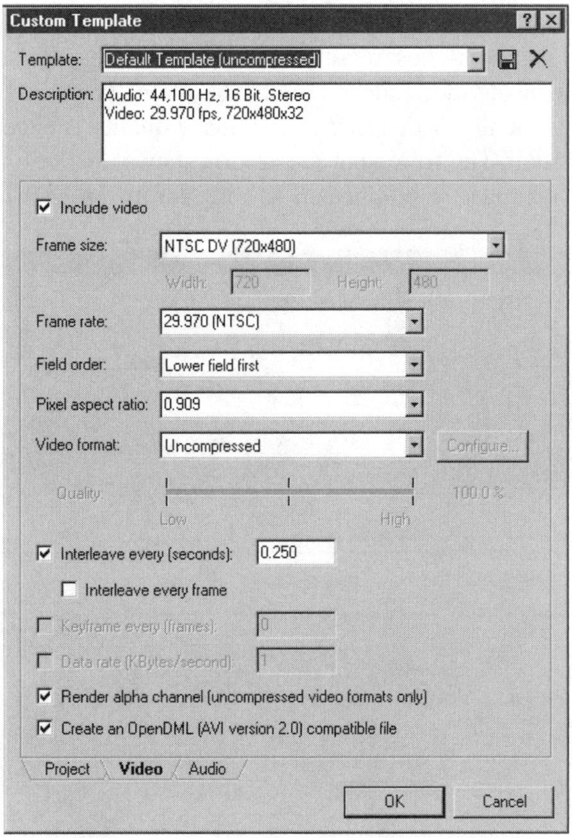

About Codecs

A codec is an algorithm that compresses and decompresses audio and/or video files. When a compression codec is applied to digital audio or video, certain information is lost. The ideal compression codec removes information by merging similar adjacent colors into a uniform hue when rendering a video frame or by eliminating high frequency sounds that cannot be perceived by humans when rendering an audio file. However, some codecs go over the top and degrade a file to a point where audio or visual anomalies are present, such as distortion and random clicking and popping in audio files or phantom pixels of color appearing in video files.

If you author DVD discs for clients or work with other colleagues, you'll receive digital audio and video files for use in your DVD projects. If you have a stout nonlinear video editing application at your disposal, you can edit video files before adding them to an Adobe Encore DVD project. If you have the luxury of specifying the format in which the files are delivered to you, specify uncompressed video in the AVI format, which you can then edit. If this is not an option, request that the file be rendered in the AVI format using the highest quality codec you have installed on your system.

There is a plethora of video codecs available for the AVI format. DivX is a popular codec for broadcast quality video files as the image quality is excellent and the file size is relatively small. The following illustration shows the options available when using the DivX 5.03 codec to produce an AVI file for use in a DVD project:

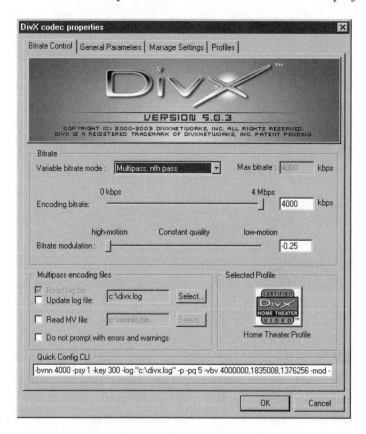

Many videographers and image editors prefer to render their video projects in the MPEG-2 Format. If your video editing application supports rendering video files in the MPEG-2 format, render the file in television standard for which you are creating the DVD. The following illustration shows settings for a video file that will be rendered in the MPEG-2 format for use in an NTSC DVD project:

There is a wide variety of video and audio codecs available. Each is suited for a specific task. Some codecs were created for the express purpose of video conferencing on the Internet and are therefore unsuitable for high quality broadcast video. Table 1-1 lists some popular video codecs with a description of what each codec has to offer. This table will help you determine if a video file compressed with a certain codec is suitable for an Adobe Encore DVD project.

NOTE *Remember, you can always choose a different codec when rendering a project from a nonlinear video editing application. However, if the original file is already heavily compressed, you will not be able to increase the quality by choosing a codec suited for broadcast quality video.*

Video Codec	Description
Indeo Video 5.1	An AVI codec commonly used for video distributed for Internet viewing on slower computers. Not suitable for broadcast quality video.
Microsoft RLE	An AVI codec that is useful when compressing videos with large areas of solid color such as cartoons, logos, and illustrations.
Microsoft Video 1	An AVI codec used for compressing analog video. As the codec supports pixel depths of 8 and 16 bits, the image quality suffers and is generally not suitable for broadcast quality video.
Intel Indeo® Video R3.2	This AVI codec is useful for rendering previously uncompressed video that will be distributed on CD-ROM discs. This codec features better image quality and faster playback than Microsoft Video 1.
Cinepak Codec by Radius	An AVI codec that is used for rendering high quality video for distribution on CD-ROM discs.
Div X 5.03	An AVI codec that can be configured for a wide variety of intended uses from web delivery to high-quality video suitable for DVD projects.
MPEG-2 Video	This compression codec is used when compressing video as an .mpg file. You determine the quality of the video by specifying the data rate. You can control the file size by choosing CBR or VBR method.
Microsoft DV	The native capture codec used by encoding method video-editing applications in the Windows format. This codec uses a minimum of compression, offers excellent video quality, but takes up a huge amount of hard drive space.
DV	A QuickTime codec used by certain video-capture hardware. Compression is kept to a minimum, which ensures high image quality but a fairly large file size. Apple's cross-platform video application QuickTime Pro 6 features an option to export a file in DV format. If your video-editing application supports this codec, you can import DV files, and after editing, export them in either AVI or MPEG-2 format for use in Adobe Encore DVD.
Cinepak	A QuickTime compression codec suited to create high-quality video with 24-bit color depth suitable for CD-ROM playback. You determine the quality of the resulting video by specifying the data rate at which the file is played back. As a rule, this format is not suitable for DVD content.
Sorenson 3	A QuickTime compression codec that produces high-quality video files in the QuickTime MOV format. If the Pro version of this codec is used with a high data rate and Sorenson 2-pass VBR encoding, the resulting file may be suitable for inclusion with an Adobe Encore DVD project.
WMV	The native codec used by the Windows Media Player. A WMV video file with a high data rate may be suitable for inclusion with an Adobe Encore DVD project. Movie Maker, Windows XPs video-editing application does have options for rendering WMV files with a high data rate.

TABLE 1-1   Commonly Used Video Compression Codecs

NOTE   *Adobe Encore DVD 1.0 does not support the QuickTime MOV (MOVie) format or the Windows WMV (Windows Media Player) format. Video files in these formats must be edited in a nonlinear, video-editing application and then rendered in a format supported by Adobe Encore DVD.*

Compressing video is an inexact science. The settings you used to create a crisp high-quality video for one project may not work with another project. When you edit digital video that you'll be using in a DVD project, the final result depends on the quality of the video with which you are supplied. In this regard, it's always advisable to start with uncompressed video clips, and after editing, render them as uncompressed AVI files if you have the hard disk space or as high-quality MPEG-2 files. If you receive video clips from clients or coworkers, specify that they render them using the highest quality codec available to them. Remember that you will also need a copy of the codec that was used to compress the file installed on your computer.

However, if you are working with a slower computer, uncompressed AVI files may not play back correctly in Adobe Encore DVD. If you find this is the case, invest in an application that you can use to compress the video using one of the high quality codecs mentioned previously in this chapter.

About Digital Audio Formats

When you import an AVI or MPEG-2 file with audio into Adobe Encore DVD, the audio track is imported as well. The audio track in the MPEG-2 file does not have to be transcoded prior to building your project. However, from within Adobe Encore DVD, you have to transcode the audio track for AVI video into one of three audio formats: PCM audio, Dolby Digital, or MPEG audio.

Understand the PCM Format

Many nonlinear video-editing applications offer the option of exporting a project with sound in the PCM *(Pulse Code Modulation)* format. PCM audio is uncompressed and as such offers excellent fidelity. The PCM sample rate is 48 khz with two stereo channels. From within Adobe Encore DVD, you can transcode audio tracks into the PCM format.

About Dolby Digital Sound

You can also transcode sound files in Adobe Encore DVD using the Dolby Digital format. Dolby Digital (AC-3) is the third iteration of the popular Dolby Laboratories audio-encoding algorithm. Dolby Digital as licensed for consumer audio and covers the spectrum from monophonic audio to 5.1 channel surround sound. Dolby Digital sound is played back with data rates ranging from 32 kbps to 640 kbps (128 kbps to 448 kbps in Adobe Encore DVD), which ensures good audio quality with the smallest possible file size. When you transcode audio tracks from within Adobe Encore DVD, you can specify the data rate by editing the project transcode settings. Transcoding video and audio files will be covered in Chapter 7.

When a DVD disc with Dolby Digital (AC-3) sound is played, the DVD player transforms the Dolby Digital signal into two-channel stereo audio. If the original signal was mixed with Dolby Surround Sound or Dolby 5.1, and the DVD player audio output is connected to a surround sound home entertainment system, full surround sound enhances the viewers' experience. Many video editing programs such as Sony Vegas 4.0 and Adobe Premiere Pro have the option to mix surround sound.

About MPEG Sound

When you use a video-editing application to prepare video assets for a DVD project, you can render the edited video and audio file in a format supported by Adobe Encore DVD. If your application has the option to render a project as an MPEG-2 video, the rendered video soundtrack is MPEG-2 Layer II. MPEG-2 Layer II sound supports 5.1 surround sound and is backwards compatible with MPEG-1 Layer 2 sound. Both forms of MPEG sound are sampled at 32, 44.1, or 48 kHz and support multiple data rates from 32 kbps to 384 kbps.

When you transcode audio tracks within Adobe Encore DVD and choose Main Concept MPEG Sound, the sound is rendered as an MPEG-1 Layer 2 sound at 48 kHz. You can specify the data rate to suit the project on which you are working.

Note that some older DVD players may not support MPEG-2 sound. In fact, if you build a DVD project and one or more title sets have MPEG-2 sound, Adobe Encore DVD will display a warning dialog to that effect. Although the resulting DVD may work fine, if your intended viewing audience may have older equipment, you should consider converting the sound track to PCM or AC-3 video. If your video-editing application gives you the option of rendering audio as a separate file, render the audio as a Microsoft WAVE (*.WAV) file and transcode the audio file in Adobe Encore DVD using the default Dolby Digital encoder or the PCM encoder.

Choose the Proper DVD Media

Recordable DVD discs come in many flavors. Many people think that DVD stands for Digital Video Disc, when in fact it means *Digital Versatile Disc* as DVD discs can also be used to store data and as media for high-end audio, as well as video entertainment. Your choice of media depends on the device you use to record your DVD projects and the device on which the disc will be played. Currently there

Utilize Mass Storage Devices

If you author DVD discs for a living, you know that the file sizes of uncompressed AVI files, or DV files that you capture from a digital video camcorder are huge. If you're authoring several DVD projects at once, you'll quickly gobble up much of your system's available hard drive space. If this is the case, consider investing in one or more external hard drives. If you purchase USB 2.0 external hard drives, the data transfer if reasonably fast. You can use an individual external hard drive to store the captured files from a client's project. After you've edited, trimmed and exported the compiled video in an Adobe Encore DVD–friendly format, you can wipe the hard drive clean and use if for other clients. You can also carry an external USB hard drive in your briefcase, or satchel and use it to transfer large video files from a client's computer. If the client is not local, you can ship the hard drive to the client with instructions on how to hook it to their computer and then transfer the files. After the files are transferred, the client sends the hard drive back to you and you've got everything you need to begin authoring the DVD.

are the following types of media: DVD+R, DVD-R, DVD-RAM, DVD+RW, and DVD-RW. Table 1-2 shows a list of popular manufacturers of DVD media and the type of media they produce.

About DVD+ Discs

DVD+R discs are one-write DVD discs. According to the DVD+RW Alliance, DVD+R discs are compatible with about 95 percent of current DVD players. Other

Manufacturer	DVD+R	DVD+RW	DVD-R	DVD-RW	DVD-RAM
CMC Magnetics	Yes	Yes	Yes	Yes	Yes
Fuji	No	Yes	Yes	Yes	Yes
Hewlett Packard	Yes	Yes	No	No	No
Maxell	Yes	Yes	Yes	Yes	Yes
Ritek	Yes	Yes	Yes	Yes	Yes
Sony	Yes	Yes	Yes	Yes	No
TDK	Yes	Yes	Yes	Yes	Yes
Verbatim	Yes	Yes	Yes	Yes	Yes

TABLE 1-2   Manufacturers of DVD Media

sources are not as optimistic and quote figures of 85 percent compatibility. One possible drawback is that the media may not be compatible with earlier players. The media can accommodate up to 4.7GB of video, audio, or digital data.

About DVD+ RW Discs

DVD+RW discs can be erased and rewritten. DVD+RW discs are compatible with 73 percent of current DVD players. Single-sided DVD+RW discs can hold up to 4.7GB of data and can be rewritten approximately 1,000 times, making them handy devices for testing DVD projects before building them to one-write discs.

About DVD-R Discs

DVD-R discs are one-write DVD discs. They are compatible with most computer and home entertainment center DVD players, as well as many first generation DVD players. The single-sided version of the media can hold up to 4.7GB of video, audio, or digital data.

About DVD-RW Discs

DVD-RW discs can be erased and rewritten. The format is compatible with roughly 75 percent of current DVD players and most computer DVD-ROM drives. Single-sided DVD-RW discs can hold up to 4.7GB of video, audio, or digital data. The discs can be rerecorded approximately 1,000 times and are used for data storage, as well as a test bed for previewing DVD projects on set-top DVD players before committing them to one-write discs.

About DVD-RAM Media

DVD-RAM media is supported by the DVD forum, but it is not compatible with all DVD recorders or DVD players. The format was originally created for use as data storage but is now also used for video. The format is quite popular in Japan.

Using Dual-Sided and Dual-Layer Media

Dual-sided media is often used to include two versions of a presentation on a single DVD disc. For example, you can create a DVD project with a wide-screen version of a movie on one side of the disc and have a version formatted to fill the aspect ratio of a television screen on the other side.

There are also dual-layer discs on the market. A dual-layer disc can store two layers of data. Dual-layer discs are read from the same side, and therefore can hold almost twice the data of a conventional single-side disc.

You can create projects for dual-sided and dual-layer discs in Adobe Encore DVD but cannot write them directly to disc. The techniques for creating a project for a dual-sided or dual-layer disc will be covered in Chapter 15.

About CD-R Media

While you cannot author a VCD *(Video Compact Disc)* using Adobe Encore DVD, you can create a DVD for computer-only playback using CD-R media. You can build a project on a CD-R disc by specifying the disc size when creating the project, as I outline in Chapter 4. Keep in mind that a CD-ROM holds a much smaller amount of data than a DVD disc. You can build a project for computer playback on a 650MB or 700MB CD-R disc. The video quality will be the same as a DVD project. The video quality when played back is determined by the hardware and software present on the machine on which the disc is played.

NOTE *You can also create projects for computer playback on DVD media. Viewers with DVD software on their computers will be able to view DVD projects you build to DVD media or CD-R media. When you build a project for computer playback, you can add a folder of files, such as Adobe Acrobat PDF files, that can be played back using computer software. You can add the folder of files to the project when you build the DVD as outlined in Chapter 15.*

Did you know?

Replicated DVD discs

When you purchase a DVD movie from a retail outlet, it has been commercially replicated. The replication facility uses a laser beam to record the original version of the authored DVD disc to a coated glass master. The glass master disc is then molded into a stamper. The replicated DVD discs are created from the stamper using an injection mold process that produces a polycarbonate disc. A reflective surface is applied to the polycarbonate disc, which is then subjected to a lacquering process to protect the disc. The disc label is then printed on top of the lacquer prior to packaging and distribution to retail outlets or wholesale distribution centers.

Other DVD discs you encounter may have been duplicated on standard DVD media. You can tell a replicated DVD disc from a duplicated DVD disc by examining the playing side of the disc. Replicated DVD discs have highly reflective silver surfaces, whereas duplicated DVD discs have a shiny green or blue surface.

Understand Television Broadcast Standards

When you create a DVD project, you must specify the television broadcast standard for the area in which the finished DVD will be viewed. The television broadcast standard is a set of specifications including the video frame size and frame rate. The television broadcast standard for America and Japan is NTSC *(National Television Standards Committee),* and the broadcast standard for Europe and the rest of the world is PAL *(Phase Alternating Line).*

Another detail you need to consider when creating a DVD project is the region coding. You can control the country in which your DVD projects can be played by specifying a region code. When a DVD is coded for a specific region, it will play back only on DVD Players sold in that region. Region coding is a part of the DVD specification that divides the world into eight regions. Table 1-3 shows the region codes and the corresponding geographic areas for each code.

About NTSC

The NTSC broadcast standard is a specific set of standards that enable you to create DVD discs that play back properly on television sets sold in America and Japan. The standard defines a composite video signal delivering 525 lines that refreshes at a rate of 60 interlaced half-frames per second. A composite video combines all video information into a single signal. The NTSC DVD frame size is either 704×480 pixels or 720×480 pixels with frame rates of 23.97 fps (frames per second), 24 fps, 29.97 fps, or 30 fps.

Region	Locale
Region 1	U.S., Canada, and U.S. territories
Region 2	Japan, Europe, South Africa, and Middle East (including Egypt)
Region 3	Southeast Asia and East Asia (including Hong Kong)
Region 4	Australia, New Zealand, Pacific Islands, Central America, Mexico, South America, and the Caribbean
Region 5	Eastern Europe (Former Soviet Union), Indian subcontinent, Africa, North Korea, and Mongolia
Region 6	China
Region 7	Reserved
Region 8	Special international venues (airplanes, cruise ships, and so on)

TABLE 1-3 Region Codes as Specified in the DVD Standards

About PAL

The PAL broadcast standard is applicable to DVDs that will be played back on television sets sold in Europe. The PAL standard delivers 625 lines that refresh at a rate of 50 interlaced half-frames per second. The PAL DVD frame size is either 704×576 pixels or 720×576 pixels with a frame rate of 25 fps.

Understand Aspect Ratios

The *aspect ratio* defines the ratio of the width to the height of an image or video frame. Motion pictures are filmed in the wide-screen format. When prepared for TV, they are often reformatted to fit the aspect ratio of a television screen, and visual information is lost. Video recorded for television display is shot with a different aspect ratio to match the nearly square television screen.

Many videographers choose to record their productions using their camcorder's wide-screen setting. When the resulting video is captured and played back on a standard television with a 4:3 aspect ratio, a black band appears above and below the video content instead of stretching and distorting the video to match the ratio of the TV screen.

Use the 4:3 Aspect Ratio

When you create a DVD project, you specify the 4:3 aspect ratio to match the video content to the aspect ratio of a television set. The video is 4 units wide and 3 units high, almost square. When videos with a 16:9 rate are included in a DVD production for television viewing, the video content is letterboxed unless the DVD is being viewed on a wide-screen television set. A letterboxed video can be readily identified by the black bands above and below the video that fill in the gap between the width of the wide-screen video to make up for the difference in height between the two aspect ratios. Figure 1-1 shows a frame from a video with the 4:3 aspect ratio.

Use the 16:9 Aspect Ratio

Most digital video camcorders have the option to record with the 16:9 wide-screen aspect ratio. This ratio conforms to the human visual field. The actual scene conforms to the wide-screen aspect ratio, but the video is letterboxed to the aspect ratio of the device on which the video is viewed. Of course, if the video is viewed on a wide-screen device with 16:9 aspect ration, the video is not letterboxed. The scene is recorded with 16 horizontal units and 9 vertical units. When viewed on a standard television screen, or a device with the 4:3 aspect ratio, black bands above and below the video fill in the gaps, as shown in Figure 1-2.

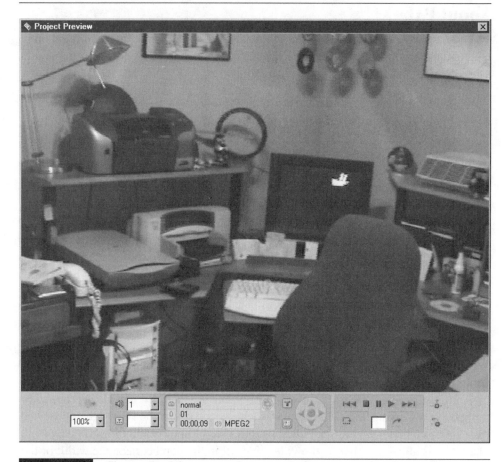

FIGURE 1-1   Choose the 4:3 aspect ratio to match the aspect ratio of a standard television set.

About Pixel Aspect Ratio

Pixels are the lowest common denominator in an image or video frame. Video pixels are rectangular, while image pixels are square. When you render a video for a specific broadcast standard, you must choose the correct aspect ratio in order for the video to display properly. Table 1-4 shows the aspect ratio for the popular broadcast standards.

About Backgrounds and Menus for Video Projects

When you use an image-editing application to create a custom background or menu for a DVD application, you must compensate for the square pixels of an image

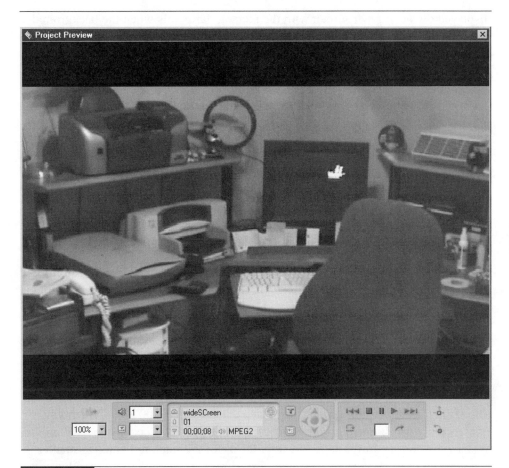

FIGURE 1-2   A wide-screen video is letterboxed when viewed on a standard television screen.

Video Type and Broadcast Standard	Pixel Aspect Ratio
Images (square pixel)	1.00
NTSC Digital Video	.9091
NTSC Widescreen Digital Video	1.2121
PAL Digital Video	1.0926
PAL Widescreen Digital Video	1.4568

TABLE 1-4   Pixel Aspect Ratios for Digital Video and Images

displayed on a computer screen with the rectangular pixels when the background is displayed on a television screen. If you do not compensate for this difference, the background or menu will not display correctly. If you create a background with a collage of images displayed over a solid color, the images will appear squashed unless you take into account the rectangular pixels of the television viewing environment. When you import a properly sized menu or background into a DVD project, Adobe Encore DVD automatically compensates for the difference in pixel aspect ratios. When your viewing audience sees the background on a television screen, it will display correctly.

If you create a background or menu in Adobe Photoshop, you can choose the proper template for the television standard for the project for which you are creating the background or menu. If you do not own Adobe Photoshop, use the appropriate size when creating a new document in your image editing application of choice. The proper size for each television standard is listed in Table 1-5.

Understand DVD Terms

In earlier sections of this chapter, you were introduced to several terms pertaining to digital video and digital audio. In this section, you'll gain an understanding of the terms that apply to the authoring of DVD discs.

About DVD Data Rate

The DVD format permits a maximum video data rate of 9,800 kbps. When you transcode a video file within Adobe Encore DVD, you can specify the data rate. When you choose a lower data rate, the file size of the rendered video is smaller, and you can fit more video on a single disc. However, when you opt for a lower data rate, the image suffers.

As mentioned previously, you can render video files using CBR *(Constant Bit Rate)* or VBR *(Variable Bit Rate)*. If you transcode video files from within Adobe Encore DVD, you can choose one of the data rate options shown in the following illustration.

Television Standard	Square Pixel Image Size
NTSC 4:3	720×534 pixels
NTSC 16:9	864×480 pixels
PAL 4:3	768×576 pixels
PAL 16:9	1024×576 pixels

TABLE 1-5   Square Pixel Image Size for Menus and Backgrounds

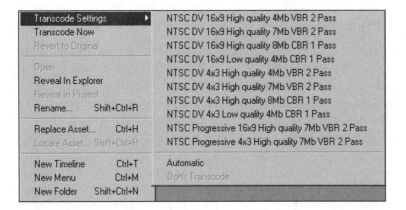

About DVD Menus

Menus are DVD navigation devices. Many DVD productions have a main menu that gives the viewer the option of viewing the entire production or navigating to a specific scene using a menu button. Menu buttons may often be linked to submenus. The DVD format is limited to a maximum of 99 title sets. You create a timeline for each video (title set) you import into Adobe Encore DVD. You'll learn how to create timelines for your DVD projects in Chapter 6.

About Chapter Points

When you create a DVD project, you can create navigation destinations known as *chapter points.* You can create a chapter point by using a project video timeline to navigate to a specific frame in the video and then use a menu command or button to create a chapter point. After you create a chapter point, you can link a menu button to the chapter point. Chapter points are often used to break large DVD productions such as full length movies into scenes. If you've ever viewed a Hollywood movie on DVD, you generally have an option to play the movie in its entirety, or play a desired scene. Each scene of the movie is designated by a chapter point.

When you create DVD projects with chapter points, your viewers can navigate to the chapter point by selecting the corresponding menu button with their DVD remote controllers and then pressing the appropriate controller button to play the chapter.

Note that chapter points must be at least 15 frames from adjacent chapter points. You are also limited to a maximum of 99 chapter points per timeline. Timelines and chapter points will be covered in detail in Chapter 6.

About Poster Frames

A *poster frame* is a frame from a video in your DVD project that is displayed on a menu button. By default, the poster frame and frame from which you create the chapter point are one and the same. However, if you are working with several video clips that have not been divided into chapter points, you can specify any frame in the video as the poster frame for the video clip. You can also specify a different frame as the poster frame when you set a chapter point in a timeline that uses multiple chapter points.

If you decide to create a DVD project with animated buttons, the button animation begins on the poster frame, and plays for the time you specify. When you create a project with animated buttons, you generally specify the buttons loop forever. Animated buttons and motion menus will be covered in Chapter 10.

About Subpictures

When you use buttons from the Adobe Encore DVD library, the buttons have subpictures. A *subpicture* is a different version of the same image that is on a different layer of the button set. When a user interacts with a menu button, a different version of the subpicture appears when the button is selected and when the menu item is activated.

Menus use color sets to determine which color is displayed when a button is selected and then activated. Figure 1-3 shows a button with the subpicture that appears when a menu button is selected. This figure was captured from a project preview in Adobe Encore DVD with the Normal button selected. Notice the highlighted border around the button. You'll learn how to work with menu color sets and buttons in Chapter 9.

You can also create text objects and convert them to buttons from within Adobe Encore DVD. You can create a subpicture for the text button from within Adobe Encore DVD. When you create a text subpicture, the button text is a different color when the user interacts with the button. The color is based on the color set you specify for the menu.

Work Within the Safe Title Area

When you create a DVD project or, for that matter, edit video in a video-editing application, you inevitably add titles and other text to your design. If you use all the available image size, parts of the title may be illegible due to the curved nature of some TV screens. The safe title area comprises 80 percent of the television screen

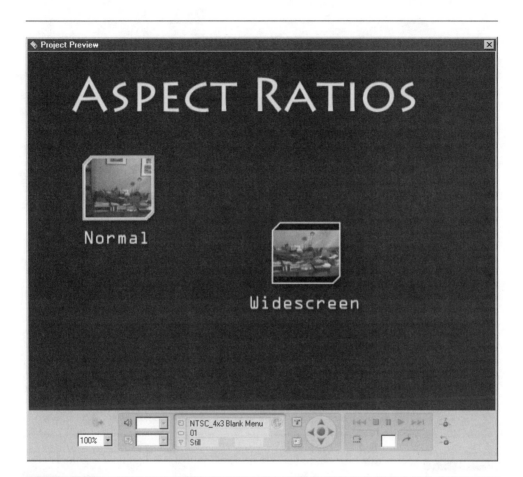

FIGURE 1-3   A subpicture appears when a viewer interacts with a DVD menu button.

measured from the center outward. Most video-editing applications give you the option to display the safe title area in the project preview window. Figure 1-4 shows a menu from an Adobe Encore DVD project with the safe title area displayed. Notice that there are two rectangles in the image. The inner rectangle represents 80 percent of the television screen (the safe title area), while the outer rectangle represents 90 percent of the television screen (the action safe area). Keep all of your title text and buttons within the safe title area and keep any important visual elements such as background images and video footage within the 5 percent safe area.

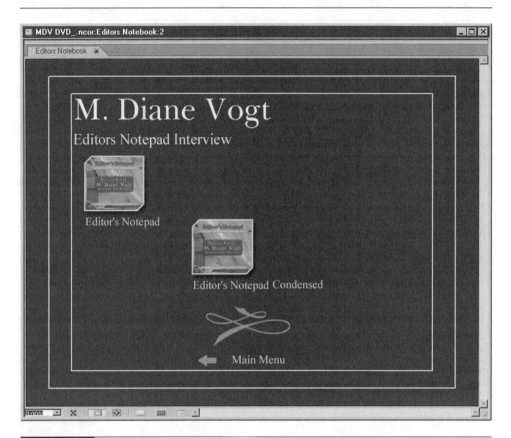

FIGURE 1-4 Keep title text within the safe title area so that it is legible when the DVD is played.

Summary

In this chapter, you were introduced to some digital video and digital audio terms and concepts. You gained an understanding of the different formats available and learned about the television standards for which a DVD project can be built. You also learned about the different media from which a DVD can be built and gained a working knowledge of the DVD format and DVD terms. In the next chapter, you'll explore the typical workflow for an Adobe Encore DVD project.

Chapter 2

Get to Know Adobe Encore DVD

How to...

- Author DVD discs
- Use Adobe Encore DVD Library presets
- Create custom DVD discs
- Explore Adobe Encore DVD workflow

There are lots of programs that can be used to author DVD discs. Many of these applications are proprietary software that ships with new computers or DVD-RW computer drives. While these DVD authoring applications can be used to create a DVD disc, they are designed for consumers with limited computer and graphic skills. As such, they offer drag-and-drop functionality and point-and-click simplicity, but they don't give the end user the features needed to create a custom disc. They also don't offer the user many options. For example, many entry-level DVD authoring programs don't give users the option to specify the action that occurs after a DVD timeline plays or the capability to add alternate sound tracks and subtitles.

If you've already started exploring Adobe Encore DVD, you know the feature-packed application offers you a great deal of flexibility. You have an impressive library of preset menus and buttons with which to work. You also have the flexibility to transcode project assets to suit the DVD project you are creating. If you create DVD discs for clients or your business, you can create custom menus and backgrounds and easily import them into Adobe Encore DVD. If you own Adobe Photoshop, you can create custom backgrounds, menus, and buttons for your DVD projects.

In this chapter, you'll gain an understanding of Adobe Encore DVD and the type of projects you can tackle with the application. You'll also explore the typical workflow of an Adobe Encore DVD project. Other topics of discussion include using the Adobe Encore DVD library presets to create menus, submenus, and buttons for your projects, importing assets into the library, and importing assets into a project. The last sections of the chapter delve into the video and audio formats that you can import into an Adobe Encore DVD project.

About Adobe Encore DVD

Adobe Encore DVD 1.0. The name says it all. It's a new application, which is actually contradictory to the word *encore*. However, a company like Adobe can indeed claim a new application is an encore, especially when you consider the depth of their product line. Adobe created Photoshop (considered by many to be the premier image-editing application on the market), Illustrator (one of the most powerful illustration applications on the market), Premiere Pro (a robust, nonlinear, video-editing application), and After Effects (an application used by professional videographers to add stunning motion graphics and visual effects to video productions). These applications are used to create graphics for the Web, print, and video. It's only logical that Adobe goes full circle and includes a DVD authoring application in its product line.

If you own any of the software mentioned in the previous paragraph, you'll be happy to know they dovetail seamlessly with Adobe Encore DVD. You can use Adobe Premiere Pro to edit and render video clips prior to importing them into an Adobe Encore DVD project. If you own Adobe After Effects, you can use the application to add visual effects to your video projects. You can also use the application to create the background videos a project's motion menus. With Adobe Photoshop you can create custom backgrounds, menus, buttons, and other graphics for your DVD projects. You can also use Adobe Photoshop to edit menus, buttons, and images in a DVD project without exiting Adobe Encore DVD. You simply select the menu you want to edit, choose the proper menu command, and a few seconds later, Adobe Photoshop launches and opens the selected menu. After editing the menu with Adobe Photoshop's powerful toolset, you save the file and your edits are applied to the project menu.

Even if the only Adobe application you own is Adobe Encore DVD, you still have one of the most powerful DVD authoring tools on the planet. In the upcoming sections, you'll begin to experience the power of Adobe Encore DVD. As you read these sections, think of ways you can use the Adobe Encore DVD features in your own work.

Author DVD Discs

You use Adobe Encore DVD to author and build DVD discs. When you use the application to it fullest, the finished product isn't your garden-variety DVD disc. When you create a DVD project with this application, you tailor the project for the television broadcast standard used in the area in which your finished DVD will be played. In addition to specifying the television broadcast standard for the DVD project, you can also specify the region in which the disc will play. If desired, you can also copy protect to disc to limit or prevent copying of the finished project. If you need to pack 50 pounds of video and audio onto a 30-pound DVD, you can transcode the audio and video assets as needed to fit the disc. Or if you decide you'd rather let the application take the reins, Adobe Encore DVD will automatically transcode project assets as needed, choosing the optimal setting that offers the best possible video quality, while still including all the assets on the specified disc size. The upcoming sections will guide you through a typical Adobe Encore DVD workflow.

Create a New Project

When a client or someone within your organization initially contacts you to author a DVD, you begin with a clean slate—a blank canvas, if you will. You take your client's or organization's vision and mold it into a finished product using your creative energy and the Adobe Encore DVD toolset. With the project visualized in your mind's eye or on paper, you launch Adobe Encore DVD and begin.

Creating a new project involves more than just launching Adobe Encore DVD and beginning to work. When you create a new project, you choose a television broadcast standard, name the disc, specify project settings, and so on. You must first organize your assets and convert them into a format that can be used within Adobe Encore DVD. You'll learn how to organize your assets, budget disc space, and much more in Chapter 4.

Import Assets

You can create a simple DVD project with a single video file or tackle a more complex project with multiple video files, still images, subtitle tracks, alternate audio tracks, and so on. You import the necessary assets into your project, and then you're ready to assemble your DVD project. You organize your assets in the Adobe Encore DVD Project tab. Figure 2-1 shows the Project tab filled with video and audio assets for a project in progress.

In Chapter 1, you learned about the digital video formats you use to create video in applications like Adobe Premiere Pro and Sony Vegas. However, Adobe Encore DVD does not support all digital video and digital audio formats. Before you can use a nonsupported file format in Adobe Encore DVD, you must first edit them in an application that supports the format, and then export it in a format that is supported in Adobe Encore DVD. In the next sections, you'll learn which video, audio, and image file formats are supported in Adobe Encore DVD.

Supported Video Formats

Video files are often the primary element in any DVD project. Whether you're building a DVD for an independent filmmaker, putting together a promotional video DVD for your corporation, or transferring your vacation videos to DVD, compelling video captures the attention of your viewing audience. Table 2-1 lists the video formats supported by Adobe Encore DVD.

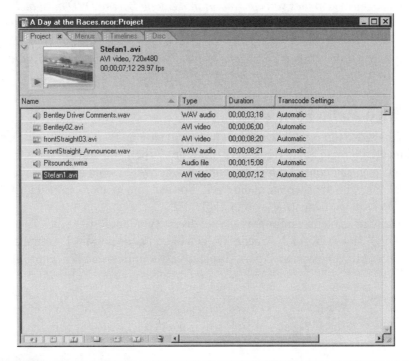

FIGURE 2-1   You import video and audio assets into a project.

Video Format	Description
AVI	This file extension is used for video files interleaved with audio.
MPG	This file extension is used for all forms of MPEG video. The file may also have an audio track.
MPEG	This file extension is also used for MPEG video.
MPE	Another variation of MPEG video. This format is often listed as *MPEG animation* or *MPEG movie*.
MPV	MPEG video stream.
M2V	MPEG-2 video-only file. This file type is commonly used for motion menu backgrounds.
M2S	MPEG-2 audio and video stream. This file type is yet another variation of the MPEG-2 format.
M2P	MPEG-2 program stream. This file type can be used for video backgrounds of motion menus.

TABLE 2-1   Video Formats Supported in Adobe Encore DVD

NOTE

The only video files you can import into Adobe Encore DVD must be NTSC or PAL compliant and sized for the DVD format. You can import NTSC-encoded video with frame sizes of 720×480 pixels, 720×486 pixels, or 704×480 pixels with a frame rate of 23.97 fps, 24 fps, 29.97 fps, or 30 fps; PAL video with frame sizes of 720×576 pixels or 704×576 pixels and a frame rate of 25 fps.

Supported Audio Formats

Most of the video files you'll be working with will have audio tracks as well. However, you can add alternate audio tracks to a timeline. When you add alternate tracks to a timeline, you can control which track plays when you create a link from a button to the timeline with the alternate audio track. You can import audio assets into Adobe Encore DVD in the formats shown in Table 2-2.

As you can see, you can import a wide diversity of audio file types into Adobe Encore DVD. However, Adobe Encore DVD will transcode non-DVD compliant audio files such as MP3, WAV, and WMA. Typically, you import audio files for alternate

Audio Format	Description
AC-3	Dolby Digital sound file. Files in this format are stereo files encoded with the Dolby Digital codec. Dolby Digital Surround 5.1 is also supported with this format.
MPEG Audio	MPEG-1 Layer 2 or MPEG-2 Layer 2 audio.
MP3	MPEG Layer 3 Audio file. MP3 files are commonly used on the Internet. When you render a sound with the MP3 format, you can achieve excellent sound fidelity with a surprisingly small file size.
MPA	MPEG Audio Stream. Files in this format may contain MPEG-1, MPEG-2, or MPEG-3 audio.
M2A	MPEG audio. Audio in files with this extension may be MPEG-1 or MPEG-2.
MP2	MPEG Layer-2 audio stream.
AIFF	Audio Interchange File Format. This sound format is commonly used by Macintosh users and can be exported from QuickTime Pro and other sound editing applications. AIFF-C is not supported in Adobe Encore DVD.
AIF	Audio Interchange File. Another sound file format commonly used by Macintosh users.
WAV	Waveform audio file. WAV files as a rule are uncompressed PCM audio with 16-bit depth and can be monophonic or stereo sound.
WMA	Windows Media Audio. This streaming audio format is used by the Windows Media Player and can be formatted as monophonic or stereo sound with a variety of bit depths and sample rates.

TABLE 2-2   Audio Formats Supported in Adobe Encore DVD

audio tracks. You'll learn all you need to know about transcoding non-DVD compliant sound files in Chapter 7. Alternate audio tracks are discussed in Chapter 12.

Supported Image Formats

Images have many uses in a DVD project. You can use images for backgrounds, menu assets, buttons, and so on. When you import images as project assets, you can display them as part of your DVD project. By default, the duration of an image on a timeline is six seconds, which can be modified. You can also use images to spice up a blank menu in a DVD project. In fact, you can use a menu command to convert an image into a fully functional menu button. Table 2-3 shows the image formats supported by Adobe Encore DVD.

Image Format	Description
BMP	Windows bitmap image format. BMP is often confused with bitmap, the term for any computer graphic. The BMP format supports 8-bit (256 colors) and 24-bit color (millions of colors) depth.
GIF	Bitmap (Compuserve). This image format is commonly used for Internet graphics and supports transparency. GIF images are limited to 8-bit color depth and are best suited to graphics with large areas of solid color.
JPG	JPEG image file (see JPEG).
JPEG	Joint Photographic Experts Group image file. JPEG is a lossy image file format. When you save a image in the JPG or JPEG format, you specify the image quality. When you reduce image quality, color information is lost, which results in a smaller file size. A highly compressed JPEG image is not suitable for a DVD project.
TGA	Targa bitmap image format. The Targa image file format supports 8-bit, 16-bit, 24-bit, and 32-bit color depth. An 8-bit color palette is comprised of 256 colors while a 16-bit color palette is comprised of 65536 colors (256×256), a 24-bit palette is comprised of 16,777,216 colors (256×256×256), and a 32-bit color palette is comprised of 16,777,216, plus an 8-bit alpha (transparency) channel.
PNG	Portable Networks Graphics This image format supports up to 32-bit color depth.
PSD	Adobe Photoshop bitmap. This native file format for Adobe Photoshop supports 32-bit color depth. You can create menus with layers for your DVD projects in Adobe Photoshop. When you save an image in the PSD format and import it as a menu asset, Adobe Encore DVD preserves the layers as created in Adobe Photoshop.
TIF	Tag Image File (see TIFF).
TIFF	Tag Image File Format. Images saved in the TIFF format can be saved with no compression, LZW compression, ZIP, or JPEG compression. You can specify image quality when you save an image in the TIFF format and choose JPEG compression. Note that Adobe Encore DVD does not support JPEG or ZIP compression.

TABLE 2-3   Image Formats Supported in Adobe Encore DVD

Create Timelines

You create timelines for image and video assets. When you create menus, you create links from menus to the timeline. When the DVD is played, users click links to play the timeline. You can also specify which frame in the timeline will be the poster frame for the asset.

If you have a long video clip, or are creating a DVD production with a single video clip such as a movie, you create a timeline for the video clip. You add chapter points to a long video clip. Chapter points are destinations that you link to menu buttons. Viewers can decide to watch the entire timeline, or they can watch a portion of the timeline by clicking the desired button, which links to the chapter point you create in Adobe Encore DVD.

If you're creating a DVD project with alternate audio tracks, you add these to a video timeline. You can have a maximum of 8 audio tracks per timeline. Figure 2-2 shows a video timeline with an audio track. Subtitles are also added to timeline tracks. You can have up to 32 subtitle tracks per timeline. Timelines will be covered in detail in Chapter 6. Audio tracks will be covered in Chapter 12, and subtitles will be covered in Chapter 13.

Transcode Assets

If you've imported DVD-compliant files into your project, you're ready to build the project after you create menus and links. However, if any of the video files are non-DVD compliant, you will have to transcode them into a format that can be read by a DVD player or computer with DVD software. When you transcode video and audio files, you specify the data rate at which the file is played back. Remember that

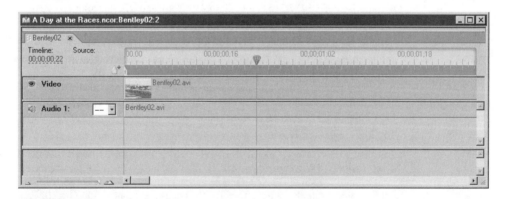

FIGURE 2-2   You create a timeline for each video asset.

the maximum data rate you can specify for DVD playback is 10,000 kbps. You can specify CBR *(Constant Bit Rate)* sampling or VBR *(Variable Bit Rate)* -2 pass sampling. When you specify VBR-2 pass sampling, Adobe Encore DVD makes a preliminary pass to review the video file for content and a second pass to determine the data rate based on the information garnered on the first pass.

You will also have to transcode any audio assets that are not DVD compliant. You can encode audio files using one of the following codecs: Dolby Digital, MainConcept MPEG Audio, or PCM audio. Transcoding video and audio assets are covered in Chapter 7.

Use Adobe Encore DVD Library Presets

When you need to quickly create a DVD with a high-quality background and top-notch menu, you can do so by using Adobe Encore DVD Library presets. The Adobe Encore DVD library contains high-quality graphic objects created by professional designers. You add a touch of professionalism by using Library backgrounds, buttons, menus, and submenus in your projects.

Use Adobe Encore DVD Menus and Submenus

Unless you're creating a DVD with a single video that will play from start to finish, every DVD project needs menus and submenus. You can choose a quality menu complete with buttons from the Adobe Encore DVD Library. Each menu has title text and button text that you can modify to suit your project. You can also choose a blank menu from the Adobe Encore DVD library, create title text using the Text tool, and then create text buttons or convert images into buttons. If you own Adobe Photoshop, you can create a menu of your own design for use in an Adobe Encore DVD project.

Create Menus with Library Presets

Adobe Encore DVD ships with an impressive array of professionally created menus and buttons, which you'll find in the Library palette. Each menu and button has default text, which you can modify to suit your project. If you own Adobe Photoshop, you can select Library presets that you've added to your project and then choose a menu command to open the asset in Adobe Photoshop. After you complete your editing in Adobe Photoshop, you save the asset, and when you return to Adobe Encore DVD, your edits are applied. Menus and buttons will be covered in detail in Part 3 of this book. Figure 2-3 shows a preset Adobe Encore DVD menu that has been modified to suit a project.

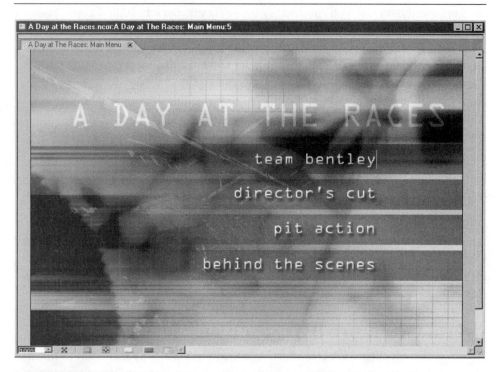

FIGURE 2-3   You can use Library preset menus for your DVD projects.

Another powerful feature of the Adobe Encore DVD library is the ability to add items such as graphics, buttons, and menus to the Library. You can add presets that you've customized in Adobe Photoshop, images that you've imported as backgrounds, and buttons to the Library. If you find that an item you've added to the Library has outlived its usefulness, you can delete it. These topics are covered in detail in Chapter 5.

Use Adobe Encore DVD Library Images

If you can't find a menu in the Adobe Encore DVD Library to suit your needs, you can choose a blank menu as the basis for your menu. You can then select an image from the Library palette for use as a background for your menu. You flesh out the menu by creating titles and menu listings with the Text tool. You convert menus listings to buttons using the Layers palette. You'll also find objects like text separators and corner graphics in the Library. Figure 2-4 shows a menu created with images and

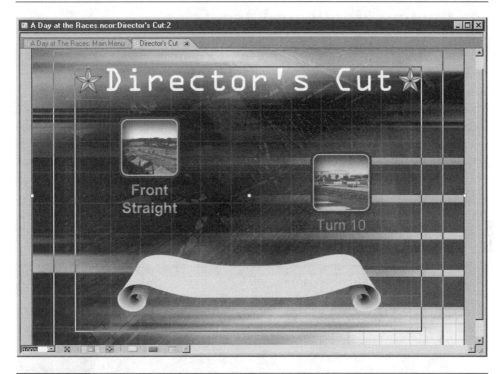

FIGURE 2-4   You can use Library graphics for your project menus.

graphics from the Library. In this image, the buttons have already been linked to video timelines, and the default button text has been modified to suit the project. Notice that the safe title area is displayed to aid in properly placing Library presets.

How to ... **Master Photoshop**

Adobe Photoshop contains many tools and options you use to create menus and buttons for your DVD projects. However, to create custom assets for your DVD project, you need to master the Photoshop toolset and learn how to work with layers. If you're new to Adobe Photoshop, you can quickly master the application by reading *How To Do Everything With Photoshop 7.0* by Laurie McCanna (McGraw-Hill/Osborne).

Create Custom Menus and Assets in Adobe Photoshop

If you own Adobe Photoshop, you can create striking menus for your DVD projects. Add several ounces of your creative energy, some sweat equity, and Adobe Photoshop's award winning toolset, and you have the basis for a compelling, eye-catching, professional menu for your DVD project. You can also use Adobe Photoshop to create custom buttons, complete with subpictures for your DVD projects. After you create the custom menu in Adobe Photoshop, you save it as a PSD file. You choose the Import As Menu Asset command in Adobe Encore DVD to bring the custom menu into your project, complete with editable text and layers. Figure 2-5 shows a custom menu being created in Adobe Photoshop.

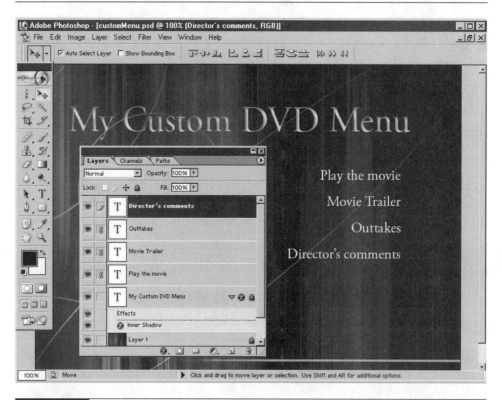

FIGURE 2-5   You can create dazzling custom menus in Photoshop.

Add Buttons to Menus

If you use one of the Library menu presets, you already have buttons for menu navigation. When you use a blank menu, you can add preset buttons for menu navigation. Most buttons on a library menu, or preset buttons you add to a blank menu, have a frame and default text. However, as soon as you link them to a timeline, the timeline poster frame appears in the button frame. If the preset menu doesn't have enough buttons for your project, you can always copy, paste, and then arrange buttons. You can also resize buttons and change to default text to suit your project. Figure 2-6 shows a typical menu with buttons that was created from a Library preset.

Create Buttons in Adobe Photoshop

If you're familiar with creating layer sets in Adobe Photoshop, you'll be able to grasp the technique of creating custom buttons in Adobe Photoshop. When you create

FIGURE 2-6   You can use a Library preset with buttons as the basis for your DVD menu.

a layer set in Adobe Photoshop, you rename the layer set, and the layers that comprise the layer set in a manner that is recognized by Adobe Encore DVD. When you create a custom button, you can define the shape for the video thumbnail, and choose the colors that are used to display the button's normal, selected, and activated states. Creating custom menus and button layer nomenclature are discussed in detail in Chapter 11.

Link Buttons

After you flesh out a menu with buttons, you link the buttons to timelines, or chapter points within timelines. Adobe Encore DVD is quite flexible giving you the option of creating the link by choosing a timeline, or timeline chapter point from a drop-down menu or by dragging from the pick whip from the link field in the Properties palette to the desired timeline, or timeline chapter point. You can also create a button and a link by dragging a timeline or timeline chapter point from the Timelines tab into a project menu.

When you have multiple buttons on a menu, Adobe Encore DVD chooses the button routing by default. However, you can modify button routing to suit your project. Buttons will be covered in detail in Chapter 9.

Set Button, Menu, and Timeline Actions

When a DVD project is played and viewers select a button to play a selection, the timeline plays in its entirety. The default end action for a timeline is stop, which in the majority of cases is undesirable as the screen turns black when the timeline stops. The only way a viewer can then return to the menu is by clicking the proper button on the DVD remote controller, an option casual DVD users may not know how to use. You can specify the end action for a timeline, transporting the viewer to another video selection, or another DVD menu. You can also set actions for menus and overrides for buttons, a topic that will be discussed in detail in Chapter 9.

Preview the Project

Before you commit your project to a DVD disc, you can preview the project. Granted, you can build the project using rewritable DVD media, but burning a DVD disc takes a good deal of time. You'll save yourself a lot of time and frustration if you preview the project in Adobe Encore DVD prior to burning the project to disc. When you preview the project in Adobe Encore DVD, you work with an interface that contains a reasonable facsimile of a DVD controller. With the controller you can test the buttons to make sure the proper video timeline plays. When you preview the project,

Did you know?

Understand Video Folders

When you build a project to a DVD disc or create a DVD folder, the disc or folder contains a folder named Video_TS, which contains pointers to the sectors on the disc that contain the audio and video streams. Adobe Encore DVD also creates a folder named Audio_TS that does not contain any files, but is needed for recognition by the DVD player. When you specify build options for a DVD master, you can copy-protect a disc to limit the number of copies that can be made. A DVD master is built to a DLT tape, which is then replicated into discs for distribution. Copy protection in Adobe Encore DVD creates a disc that can be copied one time, or not at all. After a copy of the disc has been made, the files can still be copied from the Video_TS and Audio _TS folders to a hard drive, and then copied to a DVD disc; however, the files will be scrambled to prevent playback on a DVD player.

you can test every button to be assured that the proper action occurs when a menu ends, a timeline ends, when the remote controller menu button is clicked, and so on. You can also check for broken links. Previewing a project will be covered in detail in Chapter 14.

Build the Project

After you've previewed a project and everything is working to your satisfaction, you're on the home stretch. The only task left on your To-Do list is building the disc. You have four options for building a project:

- **Make a DVD disc**   With this option, you can build the project to a DVD disc for playback on a set top DVD player, or a computer with software and hardware (a DVD ROM drive) that can play DVD discs, or build the project to a CD-R disc for playback on a computer with DVD software.

- **Make a DVD folder**   This option creates a DVD folder on your local machine, which you can use for playback on your computer. You can also use a DVD folder as the basis for burning a project to disc.

- ■ **Make a DVD image**   This option creates a DVD image on your hard drive that can be used for replicating the disc using Adobe Encore DVD or third-party software.

- ■ **Make a DVD master**   This option is used to build the project to *Digital Linear Tape* (DLT). The DLT master is used by a DVD replicating service to create multiple copies of the disc. You must have a DLT drive connected to your machine to create a DVD master. DVD masters must also used when you add copy protection to a disc.

Summary

In this chapter, you gained an understanding of the workflow involved in a typical Adobe Encore DVD project. You learned which type of assets you can import into a project and were given a preview of how you can use Adobe Photoshop to create menus and buttons for your DVD projects. In the next chapter, you'll learn to negotiate the Adobe Encore DVD workspace and how to modify it to suit your working preferences.

Chapter 3

Explore the Adobe Encore DVD Workspace

How to...

- Navigate the workspace
- Preview timelines
- Use Adobe Encore DVD palettes
- Use Adobe Encore DVD tools
- Customize the workspace

Whether you're a newcomer to Adobe software or a seasoned veteran with Adobe applications, Adobe Encore DVD is a new product; therefore, there are menu commands and tools that are not found in other applications. In this chapter, you'll gain a working knowledge of the Adobe Encore DVD workspace and the tools you use to author DVDs. You'll be familiarized with the various tabs, palettes, and menu commands you use to organize your video assets, add menus to your projects, set chapter points, and so on. You'll also learn to set Adobe Encore DVD Preferences to suit the DVD projects you create. Another topic of discussion is customizing the workspace to suit your working habits.

Introducing the Adobe Encore DVD Workspace

When you create a new project in Adobe Encore DVD and import audio assets, image assets, and video assets, you use the various tabs, palettes, and tools to create your DVD disc. The workspace has a small cast of characters: the Project window, the Properties palette, the Palette window, and the toolbar. At the top of the interface is a menu bar. When you initially start a project, all that's visible is the Project Window and any other palettes you left open when you last used Adobe Encore DVD. As you begin adding assets, you use the other interface features to perform various tasks such as adding menus to your project, linking to video assets, and so on. Figure 3-1 shows the Adobe Encore DVD workspace in action with all features visible.

Use the Project Window

You can think of the Project window as command central. From within the Project window, you can select various project assets and perform tasks such as adding menus to the project, creating timelines and so on. The Project window, shown next, is divided into four tabs: the Project tab, the Timelines tab, the Menus tab, and the Disc tab.

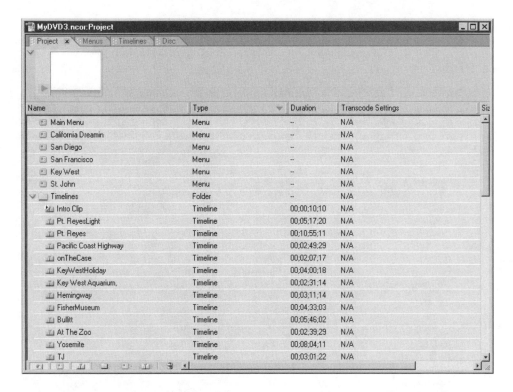

As your project takes shape, you navigate from tab to tab to add functionality to your DVD disc. For example, after importing a video asset into the Project you would

- Select a video asset in the Project tab.

- Use a menu command or button to create a timeline for the video asset.

- Switch to the Timelines tab.

- Select a timeline and add chapter points to the timeline and set chapter poster frames in the Timeline window.

In the following sections, you'll gain a working knowledge of each tab in the Project window.

About the Project Tab

The Project tab lists all the assets you've imported into a project, along with the items you create, namely menus and timelines. You use the Project tab to select assets before performing other tasks such as creating a video timeline. You also use the Project tab to manage assets and items you've created, a task that will be discussed in Chapter 5. A unique icon signifies each asset type.

Toolbar

Project window

Palette window

Properties palette

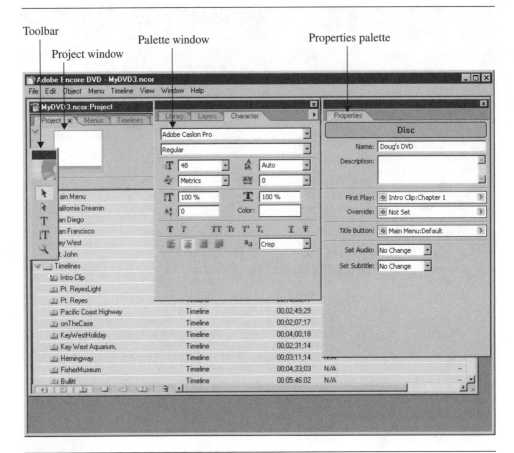

FIGURE 3-1   The Adobe Encore DVD workspace has four main elements.

The Project tab, shown next, lists each asset in your project. Note that this illustration shows the Project tab in a separate window with a video asset selected. Notice the image in the small preview window in the upper-left corner of the tab. Information about each asset is shown in columns. From within the Project tab you can:

- Display or hide project assets.

- Display or hide project menus.

- Display or hide imported assets, menus, and timelines.

- Preview a video or audio asset by selecting it and clicking the Play button in the small preview window near the upper-left corner of the Project tab.

- Preview a menu or timeline by double-clicking its name. Double-clicking a menu title opens it in the Menu editor. Double-clicking a timeline opens it in the Timeline window.

- Create Project folders.

- Delete Project assets.

- Create new menus.

- Create new timelines.

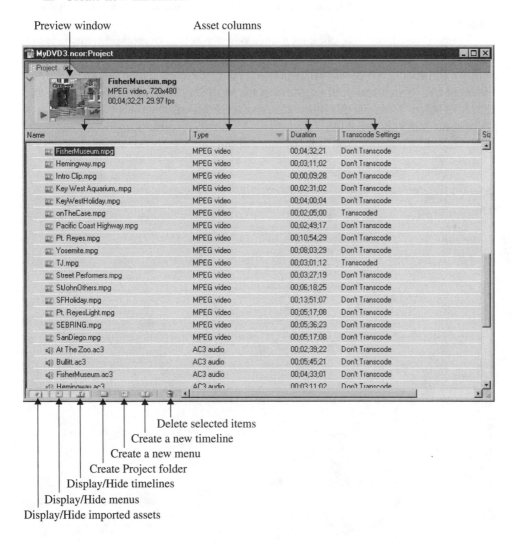

Preview window   Asset columns

Delete selected items
Create a new timeline
Create a new menu
Create Project folder
Display/Hide timelines
Display/Hide menus
Display/Hide imported assets

You can select multiple items from the Project tab. You can select contiguous items by clicking the first item and last items you want to select while holding down the SHIFT *key. You can select noncontiguous items by selecting an item and then* CTRL-*clicking additional items you want to select.*

About the Menus Tab

After you add assets to a project, you need to create some sort of navigation that viewers can use to select the chapters in your project. From within the Menus tab, you can edit menus, preview menus, rename menus, edit menus in Adobe Photoshop, and so on.

To open the Menus tab, click its name. When you first open the Menus tab, the names of all the menus and submenus in your project are displayed along with information about each menu. To display the buttons used in a menu, click its name. When you display menu buttons, they appear in the bottom half of the Menus tab, as shown below. Note that this illustration shows the Menus tab in a separate window. The Menus tab is divided into columns that contain information about menus, submenus, and buttons. You'll learn more about the Menus tab shown below in Chapter 8.

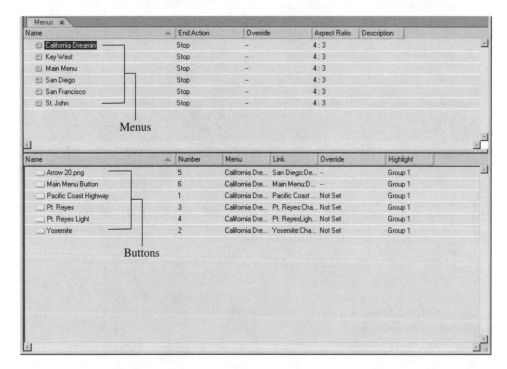

About the Menu Editor Window

You use the Menu Editor window shown in Figure 3-2 to create and edit menus in your DVD projects. From within this window you can arrange objects, add buttons, add text objects, images, and so on.

You'll receive detailed instructions on creating menus in Chapter 8. You can also preview a menu from within the Menu Editor. When you preview a menu, it opens in the Project Preview window, which contains buttons that simulate a DVD remote controller. You use these buttons to play selected DVD menu buttons, play timelines, and so on.

About the Timelines Tab

You create a timeline for each video asset in your project. To open the Timelines tab, click its title. All video timelines are displayed in the Timelines tab. When you select an individual timeline, the timeline chapter information, duration, in point, poster frame, and description are displayed in the lower half of the tab, as shown

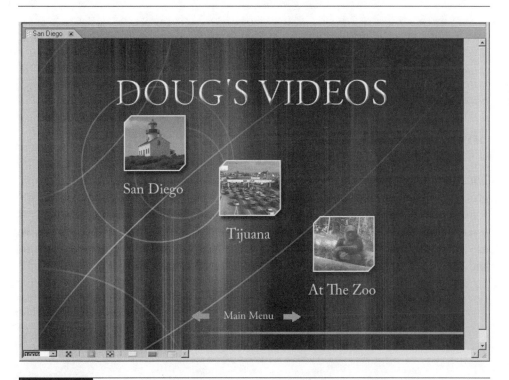

FIGURE 3-2   You use the Menu Editor to create and edit DVD menus.

in the following illustration. When you double-click a timeline, it is displayed in the Timeline window. Note that this illustration shows the tab in a separate window. You'll find detailed information on creating and working with timelines in Chapter 6.

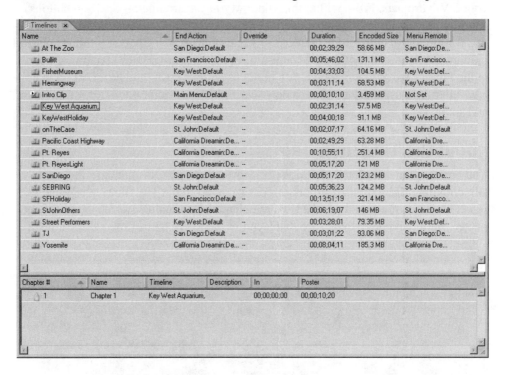

About the Disc Tab

The final tab is the Project window is the Disc tab, shown in the following illustration. You use this tab to name a project disc, specify DVD disk size, and perform other project tasks such as copy-protecting a disc. Note that this image shows the tab after a project has been created and no assets have been added to the project. As you add assets to the project, the information in this tab tells you how much disc space you've used, and how much is available. You'll find detailed information on project settings and building DVD discs in Chapters 14 and 15.

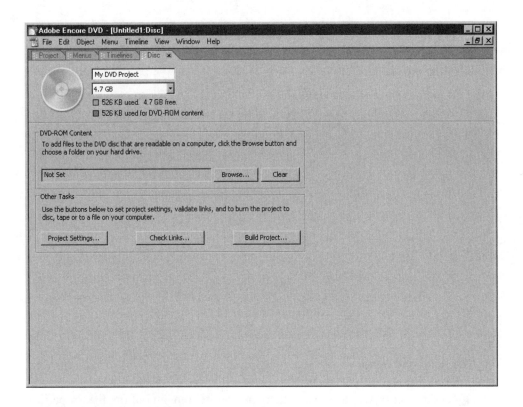

3

About the Properties Palette

Every object in a project has properties. You use the Properties palette to view an object's properties. To open the Properties palette, choose Window | Properties. The Properties palette in the following illustration shows the properties for a video asset. In fact, you can readily identify the item type for which the palette is displaying properties by reading the title at the top of the palette.

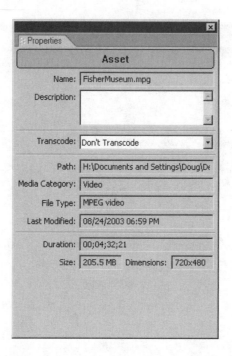

The fields and options vary depending on the type of asset or project item you choose. For example, if you choose a menu, you have options to rename the menu, specify ending actions, and so on. When you choose a menu button, the Properties palette displays fields that you use to specify the chapter point, menu, or timeline the button links to, the button's override action and so on. This powerful feature will be discussed in upcoming chapters as it relates to tasks and topics.

Use the Palette Window

The Palette window consists of three palettes: the Library palette, the Layers palette, and the Character palette. You use each palette to perform specific tasks while authoring a DVD disk. The upcoming sections will familiarize you with each palette. You'll find information on each palette as they relate to performing a specific task, for example, creating a link for a button.

About the Library Palette

You use the Library palette to add preset menus and buttons to your projects. You also use the Library palette to store the custom menus you create by modifying Adobe Encore DVD presets or the menus you create in image-editing applications (such as Adobe Photoshop) and then import into Adobe Encore DVD. To open the Library palette shown in the following image, choose Window | Library.

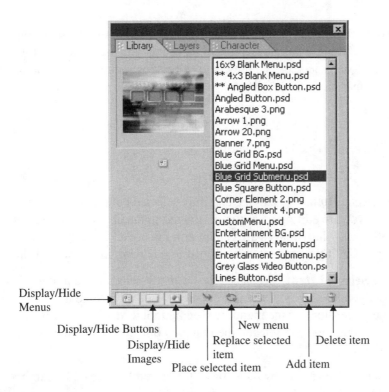

3

Display/Hide Menus

Display/Hide Buttons

Display/Hide Images

Place selected item

Replace selected item

New menu

Add item

Delete item

How to ... **Describe a Project Asset**

If you are part of a team responsible for authoring DVD discs, you'll share your projects with other team members. For example, you may be responsible for creating menu navigation while another member of your team modifies Library presets in Adobe Photoshop or creates new menus in Adobe Photoshop. In this regard, it's beneficial to have a meaningful description of each asset in a project so other team members will know the why and wherefore of an asset. You can easily add a description by selecting an asset in the Project tab, and then choosing Window | Properties to open the Properties palette. Enter a description for the asset in the Description window and then press ENTER. The description will appear in the Properties palette and will also appear on any applicable tab with the Description column displayed. For example, if you enter a description for a menu, it appears in the Project tab, the Menus tab, and the Properties palette. Adding a description to an asset will help others easily identify the asset and will serve as your personal memory jogger if you don't work on a project for a while.

After you open the Library palette, you can add a Library preset to your project. At the bottom of the Library palette is a row of icons that you use to display Library items, add presets to a project, or add presets to an existing menu. The buttons shown in the previous illustration are as follows:

- **Show Menus**   Displays or hides menus and submenus.

- **Show Buttons**   Displays or hides buttons.

- **Show Images**   Displays or hides images.

- **Place Selected Item**   Places the currently selected button or graphic within the menu you are editing. A button is aligned to the upper-left corner of the title-safe area, while images are centered in the menu. Note that this button is dimmed out if you are not currently editing a menu in the Menu Editor.

- **Replace Selected Item**   Replaces the currently selected menu item with an object you select from the Library palette. The replacement item is sized to fit the dimensions of the menu item being replaced. Note that this button is dimmed out until you select a menu item while editing the menu in the Menu Editor.

- **New Menu**   Creates a new menu using the currently selected Library menu or background.

- **Add Item**   Opens the Add Library Item dialog box that you use to locate and add a graphic to the Library.

- **Delete Item**   Deletes the currently selected object from the Library palette. This button cannot be used to delete a Library preset. You can only delete items that you have added to the Library.

Adding preset buttons and menus to a project will be discussed in detail in Chapters 8 and 9.

About the Layers Palette

When you add items to a project menu, they are arranged in layers. Items like buttons have multiple layers and are referred to as *button sets* or *layer sets*. Layer sets and button sets can be identified by the folder icon to the left of their names. Layer sets and button sets also have a right pointing triangle that is used to expand the set and reveal all layers within the set. When you work with complex menus that have multiple objects, you'll find it easier to work with individual objects by first selecting them in the Layers palette, especially when an item is a button set or layer set. To open the Layers palette shown in the following illustration, choose Window | Layers.

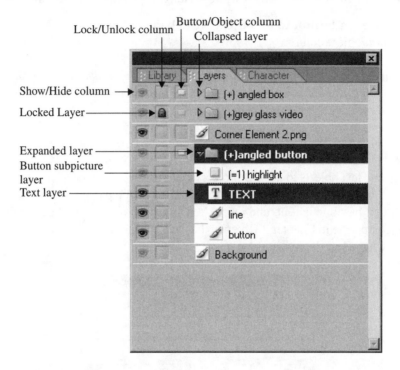

After you open the Layers palette, you can perform the following tasks:

■ **Select an item**   To select a menu item, click its name.

■ **Hide an item**   Click the eyeball icon in an item's show/hide column. When you hide an item, the eyeball icon disappears.

■ **Show an item**   Click a hidden item's display/hide column.

■ **Expand a layer set**   Click the right-pointing triangle. When you expand a layer set, the triangle points down, and all items enveloped in the layer are displayed.

■ **Collapse a layer set**   Click the down-pointing triangle.

■ **Lock a layer**   Click an unlocked layer's lock/unlock column. A lock icon designates a locked layer.

- **Unlock a layer** Click the lock in a locked layer's lock/unlock column.

- **Convert an object to a button** Click the object's button/object column. Buttons are designated by a white rectangular icon in this column.

- **Convert a button to an object** Click the button icon in the button's button/object column.

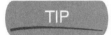

Buttons have multiple layers and have a dimmed eyeball icon in the show/hide column. To hide a button, you must first expand the layer set and then click the eyeball icon for the button graphic.

About the Character Palette

When you add text objects to a DVD project or edit text, you use the Character palette to specify font style, font color, and so on. You also use the palette to kern text, track text, shift text baseline height, and more. You'll learn how to use all the features of the Character palette in Chapter 7. To open the Character palette as shown next, choose Window | Character.

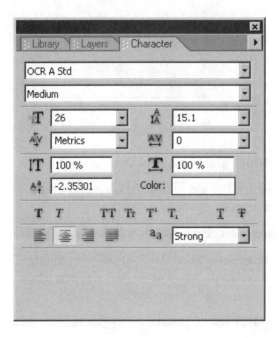

Use Adobe Encore DVD Menu Commands

At the top of the workspace is the menu bar that contains the commands you use to perform various tasks while authoring DVD projects. The menu bar is divided into groups. To choose a menu command, click a group name to display the available commands and then click the command you want to perform. Menu commands will be discussed in detail as they pertain to a topic of discussion. Many of the commands on the menu bar can also be accomplished by using the buttons you find in tabs. For example, you can use buttons in the Timelines tab to perform the same tasks as Timeline menu bar commands. The Adobe Encore DVD menu bar is divided into the following groups:

- **File**   You use the commands in this group to open projects, close projects, save projects, import assets, and so on.

- **Edit**   You use the commands in this group to undo, redo, cut, copy, paste, check menu links, and set Adobe Encore DVD preferences.

- **Object**   You use the commands in this menu group to convert objects to buttons, convert buttons to objects, create links, arrange objects, and so on.

- **Menu**   You use the commands in this group to create new menus and edit menus in Adobe Photoshop.

- **Timeline**   You use the commands in this group to create new timelines, edit timelines, add chapter points, and so on.

- **View**   You use the commands in this menu group to zoom in and out on objects, show the safe area, show subpictures, and so on.

- **Window**   You use the commands in this menu group to open palettes and to display tabs and tools.

- **Help**   You use the commands in this menu to enlist Adobe Encore DVD Help, get online support, and register the application with Adobe. There is also a command that you use for getting system information. The system information is displayed in a small window, as shown next. In addition to gathering information about your system, this command also displays all of the video and audio codecs installed on your system. Notice the Copy button at the bottom of the System Info window. Click this button to copy the system info to the clipboard. You can then paste the system info into a new word processing document, and print out a hard copy of the document. This option is useful if you ever need to contact Adobe's support team and need to share information about your system in order to track down a problem.

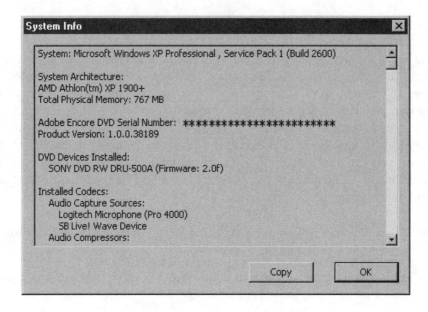

Use the Adobe Encore DVD Toolbar

You use the Adobe Encore DVD toolbar to select objects, create text, and magnify objects. You can display or hide the toolbar shown in the following illustration by choosing Window | Tools. After you display the toolbar in the workspace, you click a tool to activate it. You'll find the following tools on the toolbar:

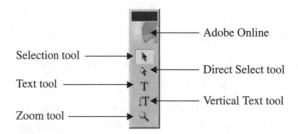

■ **Adobe Online** Click the icon that looks like a DVD disc when you're online and Adobe Encore DVD launches your web browser, which displays the Adobe Online Support Web page (www.adobe.com/support/products/encore.html).

- **Selection tool**   You use this tool to select objects or layer sets—such as buttons—that are not nested in other layer sets. After selecting an object, you can move it to a different location. You also use this tool to resize objects and select multiple objects. For more information on selecting objects, see Chapter 9.

- **Direct Select tool**   You use this tool to select objects that are nested within layer sets or button sets. For example, you'd use this tool to select a graphic, or text that is part of a button set. You can also use this tool to select items that are not in a button set. After selecting items with this tool, you can also move and resize them.

- **Text tool**   You use this tool to add text to menus and buttons. You use this tool in conjunction with the Character palette, which is used to specify font type, size, color, and so on. For more information on working with text, refer to Chapter 8.

- **Vertical Text tool**   You use this tool to add vertical text to a menu.

- **Zoom tool**   You use this tool to zoom in or out on an object. You'll find detailed instructions on using the Zoom tool in Chapter 9.

Set Adobe Encore DVD Preferences

You may use Adobe Encore DVD to author DVDs for clients, your organization, or yourself. You may create DVDs for distribution in America, Europe, or elsewhere. You can specify project settings every time you create a new project. Or if you use the same settings on a regular basis, you can save time by setting Adobe Encore DVD preferences. You can also modify Adobe Encore DVD preferences to suit your working preferences. Adobe Encore DVD preferences are divided into several categories, which are accessed through the Preferences dialog box.

Set General Preferences

When you set general preferences, you can specify the television broadcast standard that will be the default for new projects that you create. You can also specify whether tool tips are displayed when you hold your cursor over a tool or button and more. To set your general preferences, follow these steps:

1. Choose Edit | Preferences | General to open the General section of the Preferences dialog box, shown next:

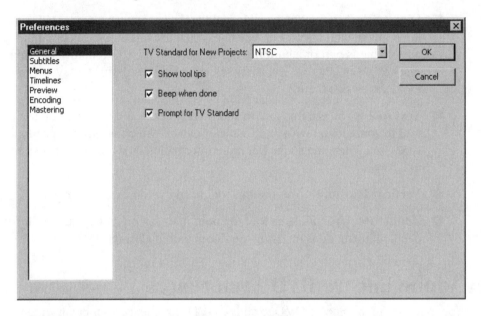

2. Click the triangle to the right of the Television Standard for New Projects field and choose NTSC or PAL. For more information on television broadcast standards, see Chapter 1.

3. Deselect the Show Tool Tips option (selected by default) to disable tool tips. If you keep the default option, a tooltip is displayed when you hold your cursor over a tool or a dialog box field for a few seconds.

4. Deselect the Beep When Done option (selected by default) if you do not want Adobe Encore DVD to beep when a build to disc has been completed. The default option also sounds a beep after a video or audio asset has been transcoded, and after a motion menu has been built. This option is handy because you can't use Adobe Encore DVD for anything else while a disc is being built, an asset is being transcoded, or a motion menu is being built.

5. Deselect the Prompt for TV Standard option (selected by default) and Adobe Encore DVD will no longer prompt you for a television broadcast standard when you create a new project. The TV standard you specify in Step 2 will automatically be applied to all new projects.

6. Click OK to apply the new preferences, or select another preference category to modify.

Set Subtitle Preferences

You can add subtitles to your Adobe Encore DVD projects. You can modify Subtitle preferences to set the default subtitle language and default subtitle length. To set subtitle preferences, follow these steps:

1. Choose Edit | Preferences | Subtitles to open the Subtitle section of the Preferences dialog box, shown next:

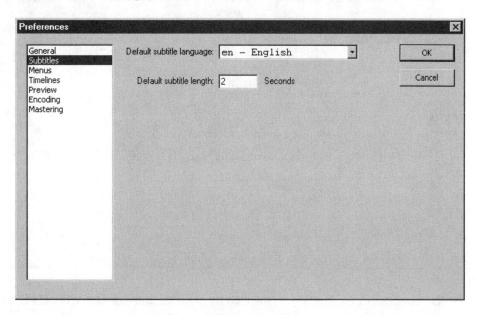

2. Click the triangle to the right of the Default Subtitle Language field and choose a subtitle language from the drop-down list.

3. Accept the default subtitle length of 2 seconds or enter a different value.

4. Click OK to apply the new settings or choose another Preference category you want to modify.

Set Menus Preferences

When you add buttons to a menu or use a preset menu from the Library, Adobe Encore DVD automatically determines the button routing unless you deselect the automatic routing when editing a menu. Adobe Encore DVD has four preset routing options from which you can choose. To modify Menus preferences, follow these steps:

1. Choose Edit | Preferences | Menus to open the Menus section of the Preferences dialog box, shown next:

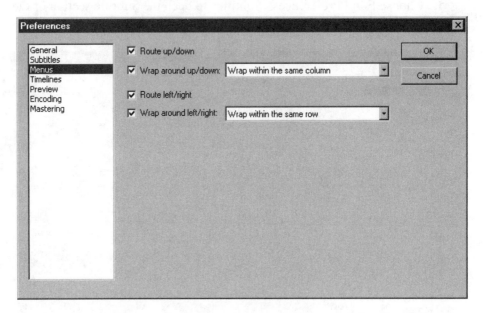

2. Deselect Route Up/Down (selected by default), and Adobe Encore DVD will not route buttons from column top to bottom.

3. Deselect the Wrap Around Up/Down option (selected by default), and Adobe Encore DVD will not wrap routing within button columns.

3

4. Click the triangle to the right of the Wrap Around Up/Down field and choose one of the following options:

■ **Wrap Within the Same Column**   Choose this option (the default), and when users click their DVD player's remote Down button at the bottom of a button column, Adobe Encore DVD will wrap to the button at the top of the column.

■ **Wrap to the Next Column**   Choose this option and when users click their DVD player remote Down button at the bottom of a button column, Adobe Encore DVD will wrap to the top button in the next column.

5. Deselect the Route Left/Right option (selected by default), and Adobe Encore DVD will not route buttons from left to right.

6. Deselect the Wrap Around Left/Right option, and Adobe Encore DVD will not wrap button routing within button rows.

7. Click the triangle to the right of the Wrap Around Left/Right field and choose one of the following options:

■ **Wrap Within Same Row**   Choose this option (the default), and when users click their DVD player remote at the end of a button row, Adobe Encore DVD wraps to the first button in the same row.

■ **Wrap to the Next Row**   Choose this option, and when users click their DVD player remote at the end of a button row, Adobe Encore DVD wraps to the first button in the next row.

8. Click OK to apply the new Menus preferences. Alternatively, select another preference category to modify.

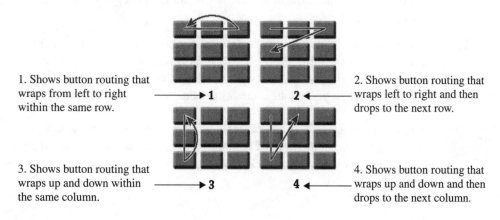

1. Shows button routing that wraps from left to right within the same row. ——→1

2. Shows button routing that 2←—— wraps left to right and then drops to the next row.

3. Shows button routing that wraps up and down within ——→3 the same column.

4. Shows button routing that 4←—— wraps up and down and then drops to the next column.

Set Timelines Preferences

If you use still images, alternate audio tracks, or subtitles in your DVD projects, you can specify the default method in which timelines handle these DVD items. To set Timelines preferences, follow these steps:

1. Choose Edit | Preferences | Timelines to open the Timelines section of the Preferences dialog box, shown here:

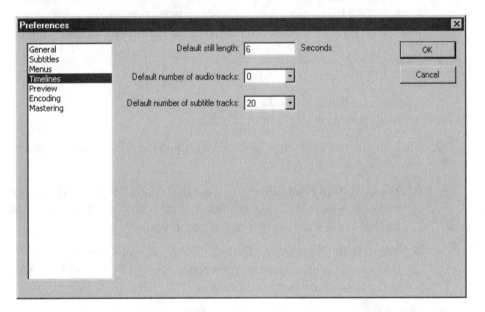

2. Accept the Default Still Length option of 6 seconds, or enter another value. This setting determines how long Adobe Encore DVD displays still images on a timeline.

3. Click the triangle to the right of the Default Number of Audio Tracks and choose an option from the drop-down list. Alternatively, you can enter a number between 0 and 8 to specify the number of audio tracks Adobe Encore DVD displays when you create a new timeline.

NOTE *If you choose a value of 0 for audio tracks and the video asset for which you are creating a timeline is encoded with an audio track, Adobe Encore DVD automatically adds an audio track to the timeline.*

4. Click the triangle to the right of the Default Number of Subtitle Tracks field and choose an option from the drop-down list. You can specify up to 32 subtitle tracks in accordance with the DVD standard; however, you cannot manually enter a two-digit value in the field.

5. Click OK to apply the Timelines preferences. Alternatively, select a different preference category to modify.

Preview Preferences

You can preview a project within Adobe Encore DVD to make sure the menus, buttons, and links are all in order. This time-saving feature enables you to catch errors before committing the project to disk. You can modify Preview preferences to suit the type of productions you create with Adobe Encore DVD. To modify Preview preferences, follow these steps:

1. Choose Edit | Preferences | Preview to open the Preview section of the Preferences dialog box, as shown here:

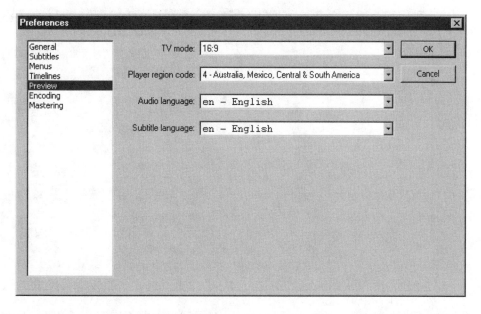

2. Click the triangle to the right of the TV Mode field and choose 4:3 Letter Box or 16:9. This option configures the Adobe Encore DVD preview player to standard or widescreen mode.

3. Click the triangle to the right of the Player Region Code field and choose the code for the region in which your DVD projects will be played back.

4. Click the triangle to the right of the Audio Language field and choose the desired language for your audio tracks from the drop-down list.

5. Click the triangle to the right of the Subtitle Language field and choose the desired language for your subtitle tracks from the drop-down list.

6. Click OK to apply the preference settings. Alternatively, you can select a different preference category to modify.

Set Encoding Preferences

When you import audio tracks that are non-DVD compliant into a project, you must encode them prior to building your DVD project. You can specify the default encoding format by setting Encoding Preferences. To set Encoding preferences, follow these steps:

1. Choose Edit | Preferences | Encoding to open the Encoding section of the Preferences dialog box, shown next:

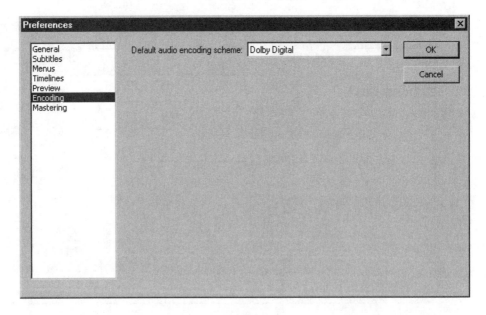

2. Click the triangle to the right of the Default Audio Encoding Scheme field and choose one of the following: Dolby Digital, MPEG-1 Layer 2, or PCM. For more information on audio encoding, refer to Chapter 1.

3. Click OK to apply the new Encoding preference. Alternatively, select another preference category to modify.

Set Mastering Preferences

When you create a DVD master that will be used to replicate hundreds or thousands of copies of a DVD project, you can specify default copy protection. When you apply copy protection to a DVD, you limit the ability to copy the disc content. Copy protection is disabled by default. You can specify copy protection for your DVD projects by setting Mastering preferences as follows:

3

1. Choose Edit | Preferences | Mastering to open the Mastering section of the Preferences dialog box, shown here:

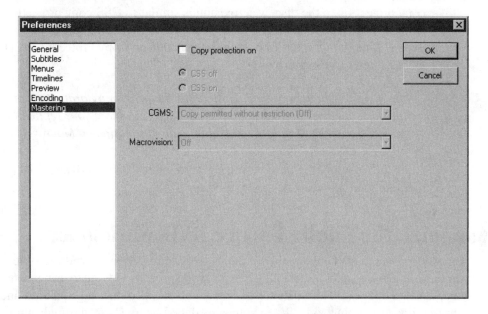

2. Select the Copy Protection On option to enable the copy protection options.

3. Choose one of the following CSS *(Content Scrambling System)* options:

   ■ **CSS On**   This option creates a CSS decryption key that can be read from the original disc only, which prohibits a digital copy from being played back.

   ■ **CSS Off**   This option does not include a decryption key on the original disc.

4. Click the triangle to the right of the CGMS *(Copy Generation Management System)* field and choose one of the following options:

    ■ **Copy Permitted Without Restriction (Off)**   This option allows unlimited copies of the disc to be created.

    ■ **One Generation Allowed**   This option permits one copy of the disc to be created.

    ■ **No Copies Allowed**   This option prohibits copying of the disc.

5. Click the triangle to the right of the Macrovision field to apply Macrovision copy protection to the disc. Macrovision is similar to CSS copy protection. Choose one of the following from the drop down list: None (the default), Type I, Type II, or Type III. For more information on Copy Protecting DVD projects, refer to Chapter 14.

NOTE   *When discs to which you have applied Macrovision copy protection are replicated, you must pay a per-disc licensing fee. Contact your replication service center for current rates and additional information on Macrovision.*

6. Click OK to apply the new Mastering preferences. Alternatively, select a different preference category to modify.

Customize the Adobe Encore DVD Workspace

The Adobe Encore DVD team laid out the workspace to suit most working styles. They also engineered a lot of flexibility into the workspace. In addition to using menu commands to open and close palettes and tabs, you can also rearrange palettes and tabs to suit your working style. When you exit Adobe Encore DVD, the application remembers the position in which you last left the workspace and restores your layout the next time you launch the application.

You have a couple choices: you can rearrange the order in which palettes appear within their window, or you can move a palette into its own window.

To rearrange the order in which tabs or palettes appear within their window, follow these steps:

1. Open the window or palette whose tab order you want to rearrange.

2. Click the title of the tab or palette you want to move and drag it to the desired position within the window.

To display a tab or palette within its own window, follow these steps:

1. Open the window or palette that contains the palette or tab you want to display in a separate window.

2. Click the title of the desired tab and drag it from the window.

If you prefer, you can also group palettes. For example, you can group the Layers palette with the Properties palette. You cannot, however, group tabs with palettes. To group palettes, follow these steps:

1. Open the window that contains the palette you want to combine with another palette.

2. Follow the previous steps to display the palettes you want to group in their own windows.

3. Click the title of one palette and drag it into the other palette. The following illustration shows a group comprised of the Properties, Character, Library, and Layers palettes:

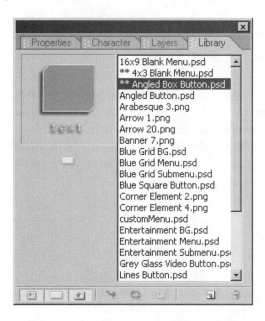

TIP   *If after rearranging tabs and palettes, you find the new setup doesn't suit your needs, you can restore palettes and tabs to their default locations by choosing Window | Reset Palette Locations.*

Use Shortcut Menus

Many software users prefer the ease of selecting commands from shortcut menus.
If you fall into this group, you'll be happy to know the Adobe Encore DVD design
team has liberally sprinkled shortcut menus throughout the application. The commands
you find on shortcut menus pertain to the tab or palette in which you are working,
or to the object you have selected. To access a shortcut menu select the object, palette
or tab in which you need to perform a command and then right-click. The following
is an illustration of the shortcut menu you access when right-clicking a menu button
in the Menu Editor:

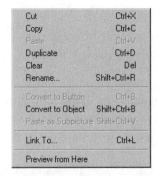

Cut	Ctrl+X
Copy	Ctrl+C
Paste	Ctrl+V
Duplicate	Ctrl+D
Clear	Del
Rename...	Shift+Ctrl+R
Convert to Button	Ctrl+B
Convert to Object	Shift+Ctrl+B
Paste as Subpicture	Shift+Ctrl+V
Link To...	Ctrl+L
Preview from Here	

Summary

In this chapter, you gained a working knowledge of the Adobe Encore DVD
workspace. You learned how to access the various tabs and palettes you use to author
DVD discs, how to modify the tabs and palettes to suit your working needs, and how
to access Adobe Encore DVD tools. You also learned to modify preferences to suit
your working needs and the type of DVD projects you create. In the next chapter,
you'll learn how to set up a DVD project.

Part II

Author a DVD

Chapter 4

Create a Project

How to...

■ Plan your project

■ Import PSD files as menus

■ Organize assets

■ Archive assets

The first step on the path to authoring a DVD is creating a new project. When you create a new project, you specify project settings, choose disc media, and so on. You determine the project settings based on the television broadcast standard for the area in which your DVD project will be played. When you set up your project, you also specify disc settings, which enables you to keep track of the amount of disc space you use as you add assets to the project.

In this chapter, you'll learn how to create a project as well as how to perform the preliminary tasks for authoring a DVD, such as planning your project, preparing your assets, and creating the actual project. You'll also learn to save your project so that you can access and edit the project at a later date.

Plan Your Project

Whether you are creating a DVD for a client, your company, or to display your own talents as a videographer, a bit of initial planning will save you time when you're deep into the process of authoring a DVD. When you're planning your project, you take into account the video and graphic assets you're using to create the DVD. You also need to determine the type of content you're authoring to DVD. Finally, you need to consider the viewing audience. Your goal as a DVD author is to create a compelling package with easy-to-navigate menus. However, the menus should be appropriate for the DVD content and your intended viewing audience, of course. As a rule, you'd use different background graphics for an educational DVD than you would for an entertainment DVD.

If your client wants a state-of-the-art DVD, you can use a background video to create a motion menu. If this is the case, you must also consider the type of background video that will be used. You should strive for a background video that is visually compelling, yet one that will not distract from title text or menu buttons. Another thing to consider when adding a motion menu to a project is the length of the video. An exceptionally long background video will take up a lot of room on the final DVD disc. If your client wants animated buttons, it's a good idea to stick with a static background; a video background and animated buttons may mix like oil and water.

DVD Checklist

It makes good sense to plan. It's also good practice to make sure you've got everything ready before starting—the actual items you need will vary from project to project. Answers to the following questions will help you ascertain what needs to be done before and after you create a project:

- How many video clips will be displayed on the DVD?

- How many chapter points will be needed for each video clip?

- Are the project assets DVD-compliant?

- If the video assets are not DVD-compliant, can you obtain the original digital video footage as captured from film or digital tape?

- Are background images sized properly for the project's television broadcast standard?

- What size media will be needed for the project?

- Will the project assets need to be transcoded in order to fit the project media?

- Does the client require copy protection?

- Will the client provide DVD-compliant assets?

- Will the client's production team be involved in the creation of the DVD?

- Will the client's production team provide graphics for the DVD?

- Is the project designed to entertain or inform the intended viewing audience?

The answers to these questions will go a long way in determining the steps you'll take to create the DVD. If you're working for a client, the answers to these questions will be major factors in ascertaining your fee for services. You'll also know how many menus your project needs and whether or not you'll have to prepare the assets prior to beginning the authoring process in Adobe Encore DVD.

Preview the Project Assets

If you're creating a DVD to display video and graphic assets you've created, you're already familiar with the content. But if you're authoring a DVD for your company or a client, you won't have an inkling of the type of menus you'll need, how many menus you'll need, what type of background graphics to employ, and so on until you're familiar with the project assets.

When you preview the video clips for your project, you'll see logical spots to create chapter points along with which areas of the videos needed to be trimmed. If your client or another member of your team has already edited the video clips, you can note the timecodes for chapter points while previewing the clips in your video-editing application. You can easily navigate to a timecode after you create a timeline for the video clip in Adobe Encore DVD, a technique that will be covered in Chapter 6.

Create a Project Storyboard

After you've previewed the assets for your project, the creative juices start to flow and you begin to see the end result in your mind's eye. At this point in time, you can begin planning the project in earnest. Your most important goal as a DVD author is to create simple-to-use, yet visually exciting, navigation. With the video and graphic assets of your project fresh in your mind, you can begin sketching your ideas on paper. The storyboard need not be elaborate; you can use a legal pad to create a mockup of each menu. If you're creating a DVD for video with a 4:3 aspect ratio, an 8 ½×11-inch legal pad is ideal as it is close to the same aspect ratio. If you're creating a DVD for video with a 16:9 aspect ratio, an 8 ½×14-inch legal pad is close to the widescreen aspect ratio. Draw your menu sketches in landscape mode and remember to include a border for the action safe and title safe areas. You can then start experimenting with button placement, graphic placement, and so on. If you're creating the menu from scratch in Adobe Photoshop, you can create sketches for button shapes.

When you're planning your project, you may also find it helpful to create a sketch of the menu navigation, showing which button links to which timeline, and what happens after the timeline plays. The sketch will also aid you in naming these project items as you create them. Creating a sketch of project navigation is often the first step to safeguarding against orphaned menus or broken links.

The following illustration shows a storyboard sketch of a simple DVD project that consists of two menus and two timelines. On the main menu there are three buttons. The first button is the default button, which is selected when the DVD first loads. If a viewer activates the button, the Wine Rally timeline plays. When the timeline finishes playing, the timeline's end action links back to the default button on the main menu. The second button links to the Victory Dinner timeline. When a viewer activates this button, the Victory Dinner timeline plays. When the timeline finishes playing, the end action links back to the default button. The set up button links to the default button (in this case, 5.1 Surround Sound) in the Setup menu, which is a submenu. If the viewer decides to choose 5.1 Surround Sound, the button links back to itself so that, if desired, the viewer can make a subtitle choice. If the viewer chooses a subtitle, the subtitle button also links back to the default 5.1 Surround Sound button in case the viewer decides to choose a different sound or subtitle option. When the viewer finishes choosing setup options, he or she clicks the Main Menu button, which links back to the default button on the Main Menu. Even though this is a relatively simple DVD project, you can see the potential for confusion when setting up button links. Sketching a storyboard helps simplify the process of creating menu button links and timeline end actions. Timelines are covered in Chapter 6. Working with submenus and buttons is covered in Chapter 9. Working with alternate audio tracks is covered in Chapter 12 while subtitle tracks are covered in Chapter 13.

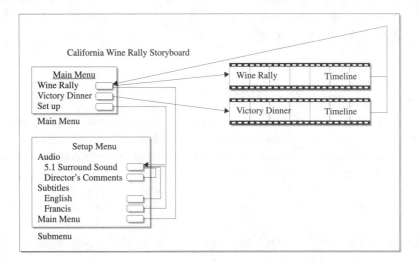

Prepare Your Assets

After previewing the project assets and sketching your storyboard, you're ready to begin preparing your assets. The amount of preparation you have to do depends on the format in which the assets are submitted to you. If you've been presented with a DV (digital video) cassette or raw footage, you'll have to capture the footage to your PC. If the footage you are working with is 16 mm or 35 mm film, you'll have to optically scan the film and digitize it to a DVD compliant format.

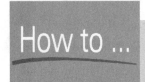

Shoot Video for DVDs (Action Safe Area)

If you've rendered any video to a DVD-compliant format and burned it to a DVD disc, you know that what you see in the camera lens isn't always what you get on the television screen. This is due to the action safe area. Remember that any images outside of the 90 percent action safe area may not be visible after the video clip is burned to a DVD disc and played back through a set-top DVD player. You don't have any control over this if you author a DVD using video clips supplied by others. However, if you shoot your own video, keep the 90 percent action safe area in the back of your mind when you compose a scene through the viewfinder. Imagine a small rectangle inside the viewfinder and don't zoom in too tightly on your subject. Strive to keep all important video information within the 90 percent action safe area, and you won't inadvertently lose the top of someone's head when you include the footage in a DVD project.

Capturing Video for DVDs

If you're a master-of-all-trades and do everything from shooting video to authoring the DVD, you'll have to capture the video from either a DV cassette or film into a digital format. The process varies depending on the application you use to capture the video. Some applications give you considerable leeway during the capture process, letting you specify the video format, frame size, frame rate and so on. If your capture application lets you specify the format and frame rate, use the following settings:

- **Video for NTSC DVDs** DV AVI video at 29.97 fps with a frame size of 720×480 pixels or 704×480 pixels

- **Video for PAL DVDs** DV AVI video at 25 fps with a frame size of 720×576 pixels or 704×576 pixels

If your video capture software gives you the option to divide the video into clips, by all means choose it. When you choose this option, the video capture software cuts the footage into individual clips that coincide with the frames where the videographer began and stopped recording a scene. Working with individual clips that are several seconds to several minutes long is much easier than working with one uninterrupted video clip.

Rendering Video and Audio for DVDs

After capturing video clips to your PC, you edit the clips and compile your video presentation. If you have total control over the DVD authoring process from start to end, you'll be editing DV-compliant video in the AVI format. If a client presents you with clips that need editing, you can trim the clips in Adobe Encore DVD; however, you have much better control over the process when you use a video-editing application such as Adobe Premiere Pro or Sony Vegas 4.0.

While you're editing clips in your video-editing application, you can also add artistic touches when transitioning from one clip to the next. If you're editing your own video, the sky is the limit. However, when you're using video transitions, you're better off if you don't mix too many transitions—they can paint a confusing picture to your viewers. Use similar transitions that are suited to the video clips. The classic cross-fade is always in good taste. If you're editing the video for a client, it's a good idea to put together a sample clip that shows the transitions you intend to use. You can render this in a smaller format of 320×240 pixels and apply sufficient compression so that it can be transmitted via e-mail, yet not enough compression to degrade the video transitions.

After editing your video clips, save them as DVD-compliant files for the television broadcast standard to which your DVD will be authored. Many video-editing applications give you rendering templates with preset options for NTSC DV and PAL DV. If your application gives you the option of specifying frame rates, choose 29.97 fps for NTSC and 25 fps for PAL.

NOTE *If you edit your video clips in Adobe Premiere Pro, you can export your clips as Microsoft AVI video and specify which codec will be used to compress the video, or you can choose Microsoft DV AVI, in which case the video is compressed using the Microsoft DV codec. Microsoft DV AVI is usually the best choice for footage destined for a DVD. You can also use the Adobe Media Encoder to export the video as MPEG-2 DVD video.*

When you're editing video clips, you should also pay close attention to the quality of the sound. This is especially important if you have a video-editing application that features multiple audio tracks like Adobe Premiere Pro and Sony Vegas 4.0. When you have multiple audio tracks, you can mix musical soundtracks into the project, and add voiceover tracks and special effects tracks. When you preview the video, watch it once with only the video soundtrack enabled. Is any of the sound distorted? Most video editing applications have vue meters, which provide you a visual reference of the volume of each track. If the meters go into the yellow or red, lower the volume of the track. Repeat this for any other audio tracks in the project.

After you've adjusted the volume for each track, play the video with all audio tracks enabled. Pay attention to the overall mix. Does the background audio track distract from the video soundtrack? Is the voiceover too loud? If so, lower the volume of the offending track(s) until the mix sounds just right. Preview the video once more and watch the vue meters to make sure they don't jump into the yellow or red. If they do, adjust the volume using the video-editing application's master volume control, which controls the loudness of the entire mix.

Your video editing may give you the option of exporting audio as a separate track. If you have this option, export the audio track as a WAV file. Better yet, if your video-editing application supports rendering project audio tracks as an AC-3 file, choose this option and specify a bitrate between 128 kbps and 448 kbps. As a rule, you'd choose 192 kbps if your audio track is stereo sound with Dolby Digital 2.0 noise reduction; 448 kbps if the soundtrack is Dolby 5.1 surround sound. When you import AC-3 audio encoded with these bitrates, Adobe Encore DVD does not have to transcode the asset. If you import WAV files as assets, Adobe Encore DVD will encode them using Dolby Digital sound (AC-3) unless you specify otherwise by changing transcode presets. Transcoding is covered in detail in Chapter 7. When you compile the DVD assets in Adobe Encore DVD, you combine the video and audio tracks when you create a timeline. You'll find detailed instructions for working with timelines in Chapter 6.

NOTE *Dolby Digital sound is licensed by Dolby Laboratories, Inc. If your video-editing application supports AC-3 Dolby Digital encoding, you may be required to purchase a plug-in in order to make this feature functional. The current release of Adobe Premiere Pro includes a trial version of the Minnetonka SurCode AC3 encoder. You can encode three audio tracks as Dolby Digital sound before you are required to purchase a license for the plug-in.*

Calculate Data Rate for Video Assets

If you know how many assets you are going to use in your project and the size of the disc to which the project will be built, you can calculate the required video data rate required to video clips of known durations on a disc with other assets. You can do this by transcoding a few calculations before launching a new project. Calculating

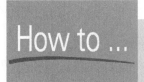

Create Chapter Points in Adobe Premiere Pro

If you use Adobe Premiere Pro as your video-editing application of choice, you can create chapter points while editing video clips. When you see a frame you'd like to use as a chapter point while previewing or editing clips in the Adobe Premiere Pro Monitor window, click the Set Unnumbered Marker button, as shown in the following illustration. Continue creating unnumbered markers at the desired frames for other chapter points in the clip. Note that adjacent chapter points must be greater than 15 frames apart.

Unnumbered Marker button

When you create unnumbered markers, they are designated on the timeline by the icons shown in the following image. You use markers as reference points in your timeline when editing in Adobe Premiere Pro. You can navigate to a marker by selecting any marker on the timeline, right-clicking, and then choosing the desired command from the shortcut menu. Note that you can also create markers on the timeline by navigating to the desired frame and then clicking the Unnumbered Marker button.

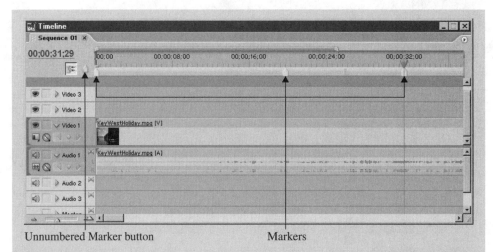

Unnumbered Marker button   Markers

You can also add information to markers in the form of comments. For a clip that will be used as a project asset in Adobe Encore DVD, you can designate a marker as a chapter point. To designate a marker as a chapter point, double-click it to open the Marker dialog box, shown next. In the Chapter field, enter the chapter name and number and then press ENTER. From within this dialog box, you can also navigate to other markers in your project and specify other chapter points. Note that you can also use this technique to create chapter points using numbered markers in your Adobe Premiere Pro project.

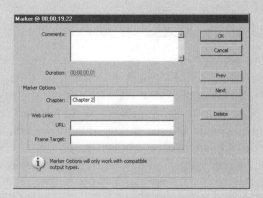

After you specify chapter points in Adobe Premiere Pro, you export the file as an MPEG file by choosing File | Export | Adobe Media Encoder and then choosing MPEG-2 DVD as the format. Choosing this export option keeps the markers you specify as chapter points intact. When you import the file into Adobe Encore DVD and create a timeline for the clip, the markers are recognized as chapter points and appear on the timeline.

the data rate for project assets is also known as *bit budgeting.* To calculate video data rate, follow these steps:

1. Calculate the available disc space in bits by multiplying the disc size by 8 (a byte = 8 bits). For example, a 4.7 GB disc is equal to 4,700,000,000 bytes × 8 = 37,600,000,000 bits or 37,600 Mbits.

2. Subtract 4 percent of the total from Step 1. This covers the overhead used by Adobe Encore DVD for DVD content and other overhead. For this example, 37,600 Mbits – 1504 Mbits = 36,096 Mbits. This is the amount of room you have available for audio, video, menus, and subtitles.

3. Calculate the space available for video assets by subtracting the size required for audio, subtitles, and motion menus from the result derived in Step 2 (36,096 Mbits). You can use the following to calculate size for these assets:

   ■ **Audio**   If you're using the Adobe Encore DVD audio encoder, Dolby Digital sound at the default data rate of 192 kbps, use a factor of 0.192 Mbps multiplied by the duration of the soundtrack in seconds. For example: two 40-minute and one 20-minute audio tracks would equal: ((2 × (40 minutes × 60 seconds) × 0.192)) + ((20 minutes × 60 seconds) × 0.192) = 1152 Mbits.

> **NOTE**   *To find the factor for other Dolby Digital Sound data rates, divide the data rate by 1,000. For example, the factor for 448 Kbps is 448/1000 or .448.*

   ■ **Subpictures**   You need not factor subpictures into your calculation unless you include subtitles. If subtitles are included, use a factor of 0.10 Mbps for each subpicture stream.

   ■ **Motion menus**   You add motion menus into the audio portion of your calculation. Motion menus have a typical data rate of 8 Mbps for each second of transcoded video. Therefore if you have one 20-second motion menu video, it uses 160 Mbits.

   ■ **Still menus**   Still menus do not factor significantly when calculating data rate, and therefore can be left out of your calculations.

For our example, 36,096 Mbits (available disc space for 4.7 GB media less 4 percent overhead) – 1152 Mbits (audio tracks) – 160 Mbps (motion menus) = 34,784 Mbits.

4. Determine the data rate required for the video in your movie by dividing the remaining disc space by the duration of the video assets (minus motion menus, which are included in the audio calculation) in seconds. In our example, there are video clips to accompany the audio clips. The resulting calculation is: 34,784 Mbits/ $(2 \times (40 \text{ minutes} \times 60 \text{ seconds}) + (20 \text{ minutes} \times 60 \text{ seconds}))$ which equates to a data rate of 5.797 Mbps, which can be rounded to 5.8 Mbps.

5. Determine the maximum video rate by subtracting the value of the audio and subtitle rates from the maximum video rate of 9.8 Mbps. In our example the calculation would be as follows: $9.8 \text{ Mbps} - (.0192 \times 3) = 9.224 \text{ Mbps}$.

As you can see in this example, we can exceed the calculated video data rate of 5.8 Mbps and achieve a higher quality video as a result. In this case we could transcode the video files with a data rate of 9.0 Mbps without exhausting the available disc space. For this project, the logical choice would be the Adobe Encore DVD High Quality 8Mb CBR 1 Pass transcode preset. If your calculations indicate you should use a data rate below 6 Mbps, be sure to use VBR (Variable Bit Rate) 2 Pass encoding.

NOTE *Keep in mind that if you've already rendered your assets to DVD-compliant formats, you should not transcode the files in Adobe Encore DVD as degradation may result.*

If all of your project assets are already DVD-compliant, Adobe Encore DVD will not have to transcode the file. When this is the case, you can approximate the required disc space adding the files sizes of the assets together.

Create a New Project

After planning your project and preparing your assets, you're ready to begin authoring your DVD. When you create a project, you specify the project settings. After creating the project, you can name the disc and specify the disc size. To create a new project, follow these steps:

1. Launch Adobe Encore DVD.

2. Choose File | New Project. The New Project settings dialog box appears.

3. Select the television broadcast standard for the project.

4. Click OK. Adobe Encore DVD displays the New Project dialog box and displays a progress bar as the transcode settings are initiated.

After you create a new project, the Project window appears along with any palettes you had open the last time you exited Adobe Encore DVD. The next step is to import assets, create timelines, and begin building navigation menus, as outlined in Chapter 5. At this stage in a project, you can also specify disc attributes if desired.

Specify Disc Size

When you create a new project, you are prompted for only the television broadcast standard for which the DVD will be built. You can, however, set other attributes from within the Disc tab. Many of these attributes do not come into play until you build the project. However, by specifying the disc size, you can monitor the remaining disc space as you add assets to your project and create timelines. To specify disc size, follow these steps:

1. Choose Window | Disc to display the Disc tab shown, next. Alternatively, you can click the Disc tab.

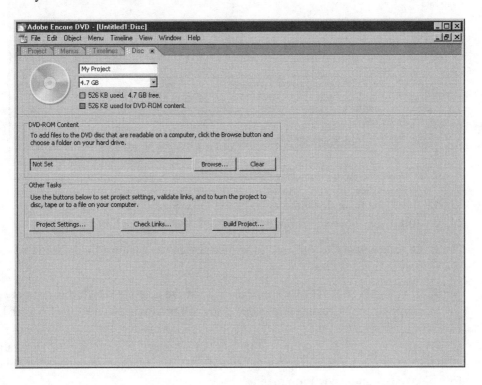

2. Enter a name in the top text field. This determines the name for the disc when you build the project.

3. Click the Project Setting button to open the Project Settings dialog box, shown next:

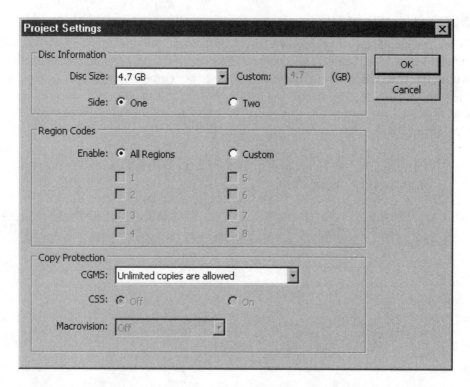

4. Click the triangle to the right of the Disc Size field and choose one of the following: 650MB, 700MB, 3.95GB, 4.7GB, 8.54GB, or Custom.

5. If you choose Custom, enter the disc size in the Custom field. Note that this size is in gigabytes.

6. If you choose 8.54GB, choose the side of the disc for which this project is being created. Creating dual-sided discs will be covered in detail in Chapter 15.

When you specify disc settings at the start of a project, you can use the Disc tab to monitor the space remaining for the specified disc size after you create timelines for video assets and navigation menus.

 *You can also specify the Region Code and copy protection for the disc using the Disc tab. For more information on region codes, see Chapter 1. For more information on copy protection, see Chapter 14.*

Check Project Size with the Disc Tab

You can keep track of the remaining disc space as you create the project. Whenever you create a new timeline, Adobe Encore DVD updates the space used and space remaining on the disc. To monitor the remaining disc space, choose Window | Disc to open the Disc tab. Alternatively, you can click the Disc tab. After you open the Disc tab, you can see how much disc space has been used by reading the information below the Disc Size window, as shown in the following illustration. The disc icon to the left of the window gives you a visual representation how much of the disc has been used and how much is remaining. As you can see in the following illustration, a large portion of disc has been used. The disc icon shows a small sliver that designates how much room is remaining.

Remaining disc space

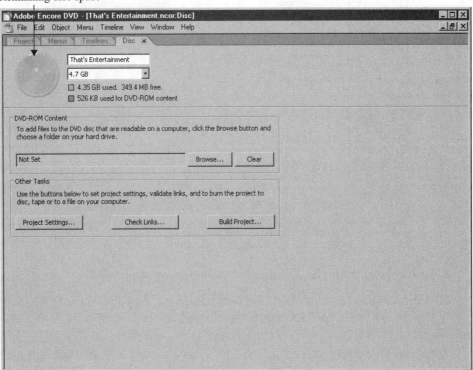

Save the Project

When you build a project, you are prompted to save it. When you save a project, Adobe Encore DVD creates a project folder, a cache folder, and a source folder. The source folder contains any assets that have been transcoded by Adobe Encore DVD and also contains a folder for menus. The project source folder contains the files as used for the project, leaving your original files unaltered. These files are stored in the Cache and Sources folder. You can save a project at any time. When you reopen the project, your menus, assets, timelines, and transcoded files appear in their proper tabs, ready for any necessary edits or additional assets needed for the project.

When you save a project, you save the file for future use, along with all the assets you import into the project. As mentioned previously, these files are saved to the project folder. To save a project, follow these steps:

1. Choose File | Save to open the Save As dialog box, shown next:

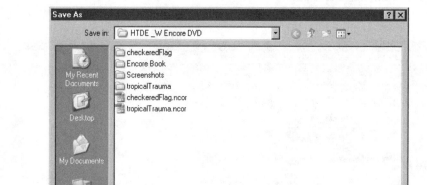

2. Navigate to the folder in which you want to save the project.

3. Enter a name for the project and then click Save. Adobe Encore DVD saves the file as an NCOR (ENCORe) file.

After you save a project for the first time, you can choose File | Save to save the file using the same filename. You can save the file as often as needed to guard against any potential computer glitches that may cause your system to crash. Adobe Encore DVD is a rock-solid application, but it uses a good deal of system resources.

When you're adding assets to the project, previewing timelines on the monitor, and so on, you may notice your system is responding a little sluggishly. If this happens, save the project and reboot your system.

When you edit menus in Adobe Photoshop from within Adobe Encore DVD, you also tax system resources. If your computer barely meets the suggested requirements for Adobe Encore DVD, it's a good idea to exit Adobe Photoshop as soon as you finish editing a menu. If you're working on a large project that spans several days, after you save the project at the end of the day, choose File | Save As and then save the project with a different filename as a backup. Computer glitches can happen at any time, not to mention power failures. If for some reason a computer glitch corrupts a project file you are working on, you can always fall back on the version you saved as a backup. Note that when you save a file under a different filename, Adobe Encore DVD duplicates everything, including the Transcodes folder, which can use up a considerable amount of disc space when you're authoring a big DVD project.

Open an Existing Project

When you work on a project with other colleagues or the project takes several days to complete, you can open the previously saved version of the project and pick up where you left off. When you open a project, Adobe Encore DVD loads transcoded files and menu files from the Project Source folder. To open a project, follow these steps:

1. Choose File | Open. Adobe Encore DVD displays the Open dialog box, shown here:

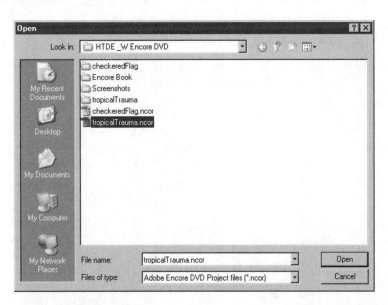

2. Navigate to the folder where your project file is saved. Remember all Adobe Encore DVD projects are saved with the NCOR extension.

3. Select the desired project and click Open. Adobe Encore DVD opens the project file.

NOTE   *Adobe Encore DVD lists the last four previously opened files at the bottom of the File command list. Choose File, then click the desired filename to open the file.*

Summary

In this chapter, you learned the first steps for authoring a DVD. You learned to plan your project, calculate video data rate, and prepare assets. You also learned how to create a new project, specify disc attributes and save a project. In the next chapter, you'll learn how to work with project assets.

Chapter 5

Manage Project Assets

How to...

■ Import project assets

■ Import PSD files as menus

■ Organize assets

■ Archive assets

After you plan a DVD project, create the project, and specify project settings, it's time to import your audio, video, and image assets into the project. After you have assembled your assets, you can use your ingenuity and creativity to author the DVD project.

You use the features of the Project tab to keep track of your assets. However, if you have a complex project with lots of assets, you can quickly become mired in organizational issues when you begin adding menus and timelines to your project. You can organize your assets to alleviate clutter in the Project tab. In this chapter, you'll learn how to import assets into a project, organize assets, and perform other tasks such as sorting and archiving assets.

Import Project Assets

After you create a project, Adobe Encore DVD presents you with a clean slate, formatted in the television broadcast standard for which your project will be played. If you've done your homework and planned the project, you have a vision of what the final project will look like when played on a DVD player. To begin fleshing out your vision, you import the audio, video, and image assets you've meticulously prepared for the project.

Import Audio, Video, and Image Assets

Your most important assets in any DVD project are audio and video. Your video files many contain audio as well. There will be other occasions where you'll be working with video only, and several audio files will be added to each timeline. A set top DVD player can only play one audio track per timeline. When you add additional audio tracks to a timeline, you specify the default track that plays, and the other tracks are alternate audio tracks. Alternate audio tracks will be covered in detail in Chapter 12.

Images can also play an important part in your DVD project. When you add an image to a timeline, it is displayed for six seconds by default. Images can be used to display product details in a DVD presentation for DVD slide shows, and can also be used for good old-fashioned eye candy. Before you can use the Adobe Encore DVD toolset to author the DVD, you import the necessary assets into the application as follows:

1. Choose File | Import as Asset to open the Import as Asset dialog box, shown next:

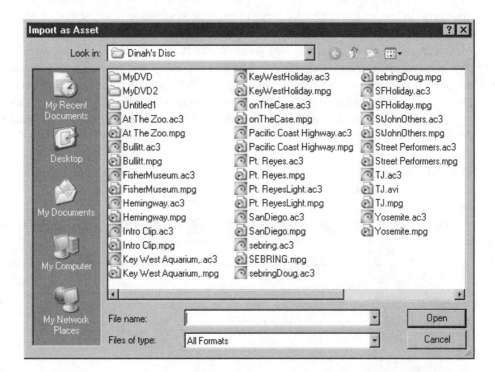

2. Accept the default All Formats for Files of Type. Alternatively, you can click the triangle to the right of the field and choose the file type you want to import from the drop-down list. Choose a specific file type if you have multiple assets in a folder and want to display only certain files, for example, AVI video files.

3. Select the file you want to import. If desired, you can select multiple files.

4. Click Open to import the asset(s), and Adobe Encore DVD imports the assets to the Project tab. This operation may take a while if you're importing multiple files or lengthy video files.

Import Menu Assets

If you own Adobe Photoshop, you can use the application to create files for use as menus. When you create a document in Adobe Photoshop, you can segregate objects on layers. When you save the document in Adobe Photoshop's native PSD format, all layers and layer sets are preserved, and objects such as text can be edited when the document is opened again. When you import a PSD file into Adobe Encore DVD as a menu asset, the layers and layer sets are preserved, and text objects remain editable. If you need to fine-tune the menu asset after importing it into Adobe Encore DVD, you select the asset, and then use a menu command to launch Adobe Photoshop.

In Chapter 11 you'll find detailed information on creating and editing menus in Adobe Photoshop. To import a menu asset into Adobe Encore DVD:

1. Choose File | Import as Menu to open the Import as Menu dialog box, shown next. Notice that the only available file type is (menu files) *psd.

2. Click the triangle to the right of the Look In field and navigate to the folder that contains the PSD file you want to import as a menu.

3. Select the file and click Open. Adobe Encore DVD imports the file into the project. The imported menu asset appears on the Project and Menu tabs and also opens in the Menu Editor window.

Scale and Swap Image Assets

When you create images for a DVD project, you generally size them to fit the aspect ratio of the television broadcast standard you specified when creating the project. If you add an image to the timeline that does not match the project aspect ratio, the image is distorted to fit the frame size, as shown here:

However, you can import images that are not sized to the project and use a menu command to scale them to the project, and then swap them to the PNG image format. After using this command, Adobe Encore DVD will add black bands to fill in the gap between the dimension that does not fit the aspect ratio and the border of the video frame. When you use this command on an image with dimensions smaller than the video frame size, the image is proportionately resized to match the video frame size, as shown in the following illustration:

You are advised not to use this command on images with dimensions smaller than the video frame as pixelation occurs when the image is enlarged. To use the Scale and Swap Asset command on an image asset, follow these steps:

1. Choose Window | Project to open the Project tab.

2. Select the image asset you want to scale and swap.

3. Choose File | Scale and Swap Asset. Adobe Encore DVD scales the image to fit the project aspect ratio and converts the image to the PNG format.

Replace Assets

After a project is underway, you may find it necessary to replace some assets with others. As an example, a member of your DVD authoring team or a client may provide you with an updated video file to use in place of a video asset currently in use. You can easily replace one asset with another as follows:

1. Choose Window | Project to open the Project tab.

2. Select the asset you want to replace.

3. Choose File | Replace Asset to access the Open dialog box, shown next:

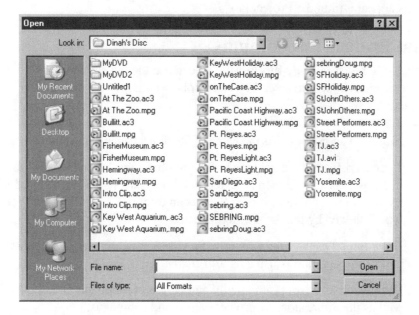

4. Navigate to the file folder in which the file you want to replace the current asset with resides.

5. Select the file and click Open. Adobe Encore DVD replaces the current asset with the file you select.

 If you try to replace an asset with different file type, Adobe Encore DVD displays the following warning dialog:

Organize Assets with the Project Tab

If you create a DVD project with a limited number of assets, the Project tab displays all of your assets for easy access. Within the Project tab, you find columns that list information about the project assets. You can display as many or as few columns to suit your needs. You can also sort assets by column and change the order in which columns appear within the Project tab.

About Project Tab Columns

The Project tab serves many functions. You can use the tab's columnar display to gain information about each asset. You can display as many of the following columns as needed:

- **Name**   This column lists the name of the project asset and cannot be hidden.

- **Type**   Displays the asset type or type of object you've created. For example, if the asset is a DVD menu created in Adobe Encore DVD, the word *Menu* appears in this column; if the object is an imported video file, the video file type followed by the word *Video* is listed.

- **Duration**   Displays the duration of a video or audio asset. If the asset is an image, the duration is listed as *Still Image.*

- **Dimensions**   Displays the width and height of video and image assets, as well as the dimensions of project menus and submenus.

- **Transcode Settings**   Lists the transcode settings for video and audio assets. If the asset is a menu or image, *N/A* (not applicable) is displayed in this column. If an asset is already DVD compliant for your NTSC or PAL project, *Don't Transcode* is displayed. For noncompliant assets, the transcode setting will be Automatic or the setting you choose for the asset. For more information on transcoding assets, see Chapter 7.

- **Size**   Displays the file size of a video, image, or audio asset. Two dashes (--) are displayed when the listing is a project folder, timeline, menu, or submenu.

- **Media Category**   Displays the type of media: Audio, Image, Video, or Video & Audio. If the listing is a project timeline, menu, or submenu, this column is blank.

- **Description**   This column is blank unless you use the Properties palette to enter a description for a project item or asset.

- **Last Modified**   Displays the date and time the project asset was last modified. If the listing is a project timeline, menu, or submenu, two dashes (--) are displayed in this column.

- **File Path**   Displays the system path to the project asset. If the asset has been moved to another system folder, and Adobe Encore DVD can't locate it, the word *Missing* is displayed in this column.

You can determine which columns are displayed in the Project tab by placing your cursor over any column and right-clicking to display the shortcut menu. Choose Columns and the name of the column you want to display or hide. Columns that are currently displayed are designated with a checkmark. Click a column name to show or hide the column in the Project tab.

You can also change the order in which columns are displayed in the Project tab. To move a column, click and drag it to a new location. As you drag the column, your cursor becomes a hand, and a rectangular bounding box appears to signify the column's current position. Release the mouse button when the column is in the desired location.

You can resize a column by moving your cursor to the boundary between two columns. When your cursor becomes two vertical lines with horizontal arrows, click and drag right to increase column width; click and drag left to decrease column width. Release the mouse button when the column is the desired width. Alternatively, you can double-click the border at the right edge of a column, which resizes the column to the length of the longest line.

TIP   *You can hide a column by placing your cursor over the column's name, right-clicking, and then choosing Hide This from the shortcut menu.*

View/Hide Project Assets

Even though you can sort project assets by type, there are times when it may be beneficial to hide certain project assets. For example, when you're working on a menu and linking buttons to timelines, you can streamline your work by hiding all assets in the Project tab except timelines. To toggle asset visibility in the Project tab, open the Project tab and do one of the following:

- ■ Click the Toggle Display of Assets icon to show or hide images, videos, and audio assets.

- ■ Click the Toggle Display of Menus icon to show or hide project menus and submenus.

- ■ Click the Toggle Display of Timelines icon to show or hide project timelines.

When items of a certain type are displayed in the Project tab, the applicable icon(s) are recessed. The following illustration shows the Project tab with the Toggle Display of Menus icon selected.

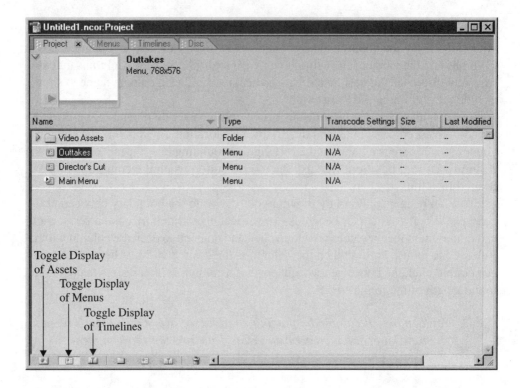

Rename Project Assets and Items

When you import assets into a project, they inherit the filename and extension of the original file. When you create a new timeline or use a menu from the Library, Adobe Encore DVD assigns the object a default name. When you work on a project of any complexity, the default names can be a hindrance; for example, when you're trying to figure out what video assets NTSC_4:3 Blank Menu 5 is for. You can alleviate any confusion about what an asset, timeline, or menu is used for when you get in the habit of giving unique names to project assets and items. To rename a project asset or item:

1. Choose Window | Project to access the Project tab.

2. Select the asset or item you want to rename.

3. Choose Edit | Rename to open a dialog box called Rename, appended by the asset type you are renaming. The following illustration shows the dialog box for renaming an MPEG video:

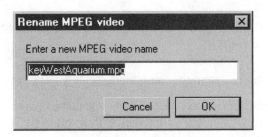

> TIP   *You can also access the Rename dialog box by selecting the asset you want to rename, right-clicking, and then choosing Rename from the shortcut menu.*

4. Enter the desired name for the asset.

5. Click OK or press ENTER to apply the new name to the asset.

> NOTE   *When you rename an imported asset, you change the name only as it appears within your DVD project. The original filename of the asset is unchanged.*

Sort Project Assets

In addition to choosing which project items and assets are displayed in the Project tab, you can also sort project items. You can sort items by column in ascending or descending order. To sort project items and assets, follow these steps:

1. Choose Window | Project to display the Project tab.

2. Click the column title by which you want to sort the assets. For example, to sort assets by name, click Name. After you sort a column, an icon appears to the right of the column name. An upright triangle designates that the column is sorted in ascending order; an inverted triangle designates descending order.

3. Click the column title again to sort the column in the opposite order.

Preview Project Assets

When you import audio, video, and image assets into a DVD project, you can preview the assets by previewing the entire project. However, this takes some time as your processor has to assemble all the menus and video assets in a format that can be displayed in the Project Preview window. When you need a quick preview of an individual asset, you can do so by using the Thumbnail Preview in the upper-left corner of the Project tab. To preview an imported asset, open the Project tab and do one of the following:

■ To preview an image in the thumbnail preview window, select it.

■ To preview an audio asset, select it and a speaker icon appears in the thumbnail preview window. Click the Play button to listen to the audio asset.

■ To preview a video asset, select it and the first frame of the video appears in the thumbnail preview window, as shown in the following illustration. Click the Play button to play the video. If the video file has embedded audio, the audio plays as well.

Play button

 *To hide the thumbnail viewer, click the inverted triangle to the left of the viewer. Click the triangle again to display the viewer.*

Organize Assets with Project Folders

When you work on a DVD project with a limited number of assets, the Project tab functions flawlessly. However, when you have a project with multiple audio, video and image assets, and so on, the Project tab quickly becomes cluttered, and it may become difficult to sift through all the assets on the list to select the desired one. You can organize the Project tab with the use of folders.

Create Project Folders

You can create folders to organize the Project tab at any time. However, if you've planned the project ahead of time and know you'll be dealing with a large number of assets, you can save yourself some frustration by creating project folders before you import the first asset. Then you can import assets into the folders you've created for them and save time when you begin authoring your DVD. To create a project folder, follow these steps:

1. Choose Window | Project.

2. Choose File | New Folder to open the New Folder Name dialog box, as shown in the following illustration. Alternatively, you can click the Create a New Folder icon at the bottom of the Project tab.

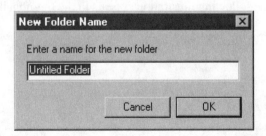

3. Enter a name for the folder and Click OK. The new folder appears as part of the Project tab list.

Add Assets to Project Folders

After you create a project folder, you're ready to add assets to the folder. When you organize a project with folders, it's a good idea to create separate folders for video assets, audio assets, and timelines. You can create another folder for miscellaneous assets as a catch-all for images and assets you add to the project from the Library.

After creating a new folder, you can:

■ Add objects to the folder by dragging and dropping them into the folder.

■ Import an asset directly into a folder by clicking the folder name and then choosing File | Import as Asset, or if you're importing a menu, choose File | Import as Menu.

■ Expand a folder by clicking the right-pointing arrow to the left of the folder's name to display the folder's contents.

■ Collapse an expanded folder by clicking the down-pointing arrow to the left of the folder's name. This hides the contents of the folder giving you more working room in the Project tab.

About Missing and Offline Files

When you add assets to a project, Adobe Encore DVD does not move the files into a folder; rather, it records the path to the asset. When you reopen a project, Adobe Encore DVD re-establishes the path to the file, and it becomes available for use in your DVD project. However, if you move a file to a different folder, the path to the file cannot be verified when you open the project. Missing or offline files are still listed in the Project tab, but they are italicized. If you select a missing or offline file, the icon shown in the following illustration appears in the thumbnail preview window:

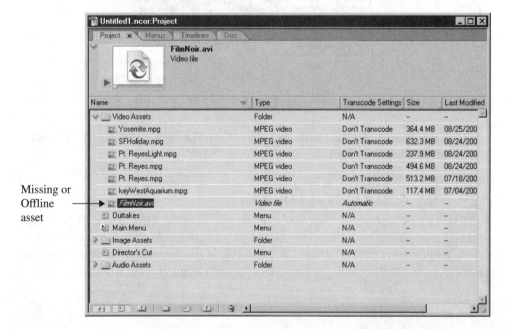

Locating Missing or Offline Files

If you open a project and discover some of the files are missing or offline, you can establish the path to the file's current location on your system or network. After re-establishing the path, the file is again available to your project. To locate a missing or offline file:

1. Select the missing or offline file in the Project tab.

2. Choose File | Locate Asset to access the Open dialog box. Adobe Encore DVD navigates to the folder in which the asset was previously stored.

3. Click the triangle to the right of the Look In field and navigate to the folder in which the asset is now stored.

4. Click Open to re-establish the link with the asset. The asset's name is no longer italicized, and the appropriate image or icon for the asset type appears in the thumbnail preview window.

Work with Library Items

As mentioned in previous chapters, the Adobe Encore DVD Library is filled with graphics, buttons, and menu items you can use in your DVD projects. The Library is a wonderful resource that you can modify to suit the DVD projects you create. You can add items to the Library, delete them, and determine which type of items are displayed. You can also add any image, menu, or button being used in a project to the Library. To open the Library shown in Figure 5-1, choose Window | Library.

Preview Library Presets

After you open the Library, you can preview Library presets to determine if they are the ideal solution for your project. Remember that you can always modify a preset menu by adding or deleting buttons or by modifying text. Modifying menus will be covered in detail in Chapter 8.

To preview a Library preset:

1. Choose Window | Library.

2. Select a preset, and a thumbnail image appears in the Library preview window, as shown in Figure 5-1. Notice the icon below the preset, which

signifies the type of item you've selected. The icons are shown in the following illustration:

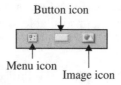

Button icon

Menu icon

Image icon

View/Hide Library Items

When you first install Adobe Encore DVD, you have a relatively easy time deciding which Library items to use in your projects. However, when you start customizing Library presets or create new items in Adobe Photoshop and then add them to the Library as new presets, the Library can become quite crowded. You can toggle

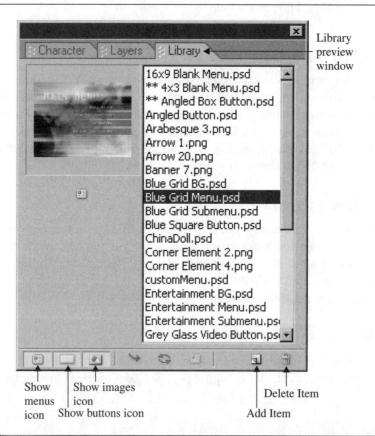

Library preview window

Show menus icon

Show images icon

Show buttons icon

Delete Item

Add Item

FIGURE 5-1   You can use preset items from the Library in your DVD projects.

the display to temporarily hide preset types you're don't currently need. To hide Library presets, do one of the following:

- Click the Show Menus icon to show or hide menu presets.

- Click the Show Buttons icon to show or hide button presets.

- Click the Show Images icon to show or hide image presets.

Import Items to the Library

When you take the time and effort to create special artwork for a menu or to create a custom menu, you've created an item you can use on other projects. You can add the item to the Library so that it's readily available for a future project. Remember that you can modify any Adobe Encore DVD Library item in Photoshop to give it a slightly different look without having to create a new item, thus streamlining your workflow and saving you time. You can import image files of the following formats into the Library: PSD, BMP, GIF, JPG, JPEG, PNG, TIF, TIFF, and EMF. To import an object to the Library, follow these steps:

1. Choose Window | Library to open the Library.

2. Click the Add Item icon shown previously in Figure 5-1 to open the New Library Item dialog box, shown next:

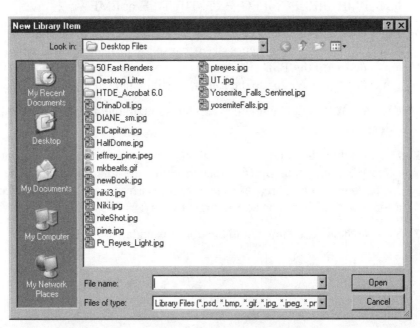

3. Click the triangle next to the Look In field and navigate to the folder in which the item you want to add to the Library is stored. You can select multiple files to add to the Library.

4. Click Open. The new items are added to the Library.

 When you add an item to the Library, it is added to the Adobe Encore DVD Library folder that is a subfolder of your Application Data folder. The Library item will continue to be a part of the Library, even if you remove the file from its original folder.

Add Assets from Projects to the Library

When you're working on a project and modify a Library preset with a tool or import a graphic, button, or menu you find especially useful, you can add it to the Library for use in future projects. When you add an image to the Library, it becomes a Photoshop PSD object. To add a project asset to the Library, follow these steps:

1. Choose Window | Project to open the Project tab.

2. Choose Window | Library to open the Library palette.

3. Select the object from the Project tab you want to add to the Library. Remember you can add objects with one of the following file types only: PSD, BMP, GIF, JPG, JPEG, PNG, TIF, TIFF, or EMF.

4. Right-click and choose Rename from the shortcut menu to open the Rename dialog box. This step is necessary because the project asset name will be appended with the PSD file extension. For example, if you add a JPG file called myImage.jpg to the folder, it will be named myImagejpg.psd.

5. Give the asset a new name or keep the same name and remove the file extension.

6. Drag and drop the object into the Library. As you drag the object beyond the Library border, a plus sign (+) appears below the cursor indicating the file can be added to the Library. If a circle with a slash appears, you have selected a file type that cannot be added to the Library.

7. Release the mouse button, and the object is added to the Library.

Delete Library Items

As you become more proficient with Adobe Encore DVD and create new items to add to the Library, you may find some of your older work pales in comparison. You can free up some real estate in the Library by deleting unwanted items. Note that you cannot delete any Adobe Encore DVD Library presets; you can delete only items you've added to the Library. To delete an item from the Library, follow these steps:

1. Choose Window | Library to open the Library.

2. Select the item you want to remove. You can select only one item at a time from the Library.

3. Click the Delete Item icon that looks like a garbage can. Adobe Encore DVD displays the Delete Library Item dialog that warns you deleting a Library item cannot be undone. If you select a Library preset, the Delete Item icon is dimmed out. Remember, you can only delete items that you've added to the Library.

4. Click OK to delete the Library item. Alternatively, press DELETE or the BACKPACE key to delete an item.

 **Archive Project Assets**

Even though Adobe Encore DVD provides you with a robust toolset for creating a professional-looking DVD, it's still time-consuming work. When you create a one-of-a-kind DVD disc for yourself, your company, or your client, you can always duplicate the disc. However, you may find it beneficial to archive all the project assets as well as the Adobe Encore DVD source file. Due to the large capacity of DVD discs, you can generally store all the project assets, the source file, and the project folder. If you organize your entire project in a single folder, launch the application you use to burn CD-ROMs and DVD-ROMs, and then follow the prompts to burn the folder to a disc. If you burn the project to a rewriteable disc, you can copy the project assets to your system when you want to modify the project. After modifying the project, you can archive the modified project on the rewriteable disc.

Summary

In this chapter, you learned to add assets to a DVD project, manage project assets, and use the Project tab. You also learned how to sort items in the Project tab, how to work with Library presets, add items to the Library, and maintain the Library by deleting unwanted items. In the next chapter, you'll learn to create timelines for your video assets, add audio tracks to the timeline, work with the Timeline window, and much more.

Chapter 6

Work with Timelines

How To...

- Create timelines
- Set first play
- Create chapter points
- Create poster frames

When you edit video in an application like Adobe Premiere Pro or Sony Vegas 4.0, you use a timeline to trim unwanted footage from clips, add video transitions and effects to clips, and compile them into a finished production. When you bring your edited video clips into Adobe Encore DVD as assets, you need to specify the points in the video clips that determine the start of each DVD chapter. You must also determine which frame in a chapter is the poster frame. In some DVD-authoring applications, you use small sliders to navigate a video clip and set chapter points and poster frames. Using a slider to navigate to a specific frame in a production can be tedious work, especially when you have a long video timeline. Fortunately, in Adobe Encore DVD you can create timelines similar to the ones you use in video-editing applications. You use Adobe Encore DVD timelines to quickly navigate to specific frames in video assets.

In this chapter, you'll learn to create video timelines and to set chapter points and poster frames. You'll also learn how to set a timeline as DVD first play, navigate timelines, and preview timelines. Finally, you'll learn to work with images on timelines, specify image duration, and much more.

About Timelines

In Adobe Encore DVD, a *timeline* is a graphical representation of a video clip. Adobe Encore DVD displays a time ruler at the top of the timeline and a graphical representation of the video timeline. When you zoom in on a timeline, many video-editing applications display thumbnail images of the frames. When you create a timeline in Adobe Encore DVD, a thumbnail of the first and last frames of the video clip are displayed. I-frames are displayed as white hash marks if you zoom in to display a span of frames in which one or more I-frames reside. In MPEG video, an I-frame designates the start of a *Group Of Pictures* (GOP). Therefore, an I-frame is also referred to as a *GOP Header.* Note that AVI file timelines do not have I-frames. The remainder of the timeline is represented by a solid blue background. Each timeline audio track is displayed as a solid line. Subtitle tracks are displayed below the audio track(s), as shown next:

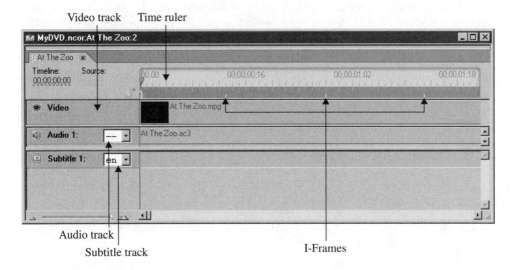

Video track   Time ruler

Audio track

Subtitle track

I-Frames

6

Each second of a timeline is divided into the frame rate of the television broadcast standard you specify when you create the project. Each second of an NTSC timeline time ruler is divided into 30 hash marks, while each second of a PAL timeline ruler is divided into 25 hash marks. When you select a frame of a timeline, its timecode is displayed as hours: minutes: seconds: and frames. As an example, a timecode of 00;00;12;25 designates that 12 seconds and 25 frames of the timeline have elapsed.

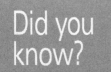

About SMPTE

SMPTE is an acronym for *Society of Motion Picture and Television Engineers;* the organization sets the standards for video and audio recording, as well as other recommended practices for creating video for motion pictures and television. Members of the organization develop standards for new technologies as needed. You can find out more about SMPTE at their web site: www.smpte.org

Create Timelines

You create a timeline for each asset that will be displayed in a DVD. A timeline can contain a single video track, multiple audio tracks, and multiple subtitle tracks. If you create a timeline for image assets, you can display as many images as needed on a single timeline. You cannot mix video and image assets on a timeline or display more than one video asset on a timeline. Note that each timeline is limited to a maximum of 99 chapter points. To create a timeline, follow these steps:

1. Choose Window | Project to open the Project tab.

2. Do one of the following:

   ■ Select a video or image asset and choose Timeline | New Timeline. Alternatively, you can press CTRL-T.

   ■ Select a video or image asset and click the Create A New Timeline icon at the bottom of the Project tab.

After you create a new timeline, it appears in the Timeline window. From within the Timeline window you can navigate the timeline, add chapter points, and specify poster frames, as outlined in the upcoming "Explore The Timeline Window" section.

 You can also use an audio asset as the basis for a timeline. When you create a timeline from an audio asset, Adobe Encore DVD provides you with a blank video track for future use. When an audio only timeline is played on a set top DVD player, the TV screen is black.

Create a Blank Timeline

If you prefer, you can create a blank timeline for future use. When you create a blank timeline, you can add video or image assets at a later date. To create a blank timeline, follow these steps:

1. Choose Window | Project to access the Project tab. Make sure you don't select any assets; otherwise, the asset will be housed in the timeline.

2. Choose Timeline | New Timeline. Alternatively, you can click the New Timeline button at the bottom of the Project tab.

After you create a blank timeline, you can add a video or image asset to the timeline as described in the "Add Assets to a Blank Timeline" section.

Create a DVD Intro Clip

To add some polish and panache to a DVD production, you can create an intro clip that plays as soon as your DVD project is inserted in a user's DVD player. The intro clip can be as simple as an image displaying a warning that the material is copyrighted, or it can be a 10 or 15 second animation that displays your client's or your company's log. Create the intro clip in a video-editing application like Adobe Premiere Pro or Sony Vegas. Add some visual eye candy by applying video effects, and then render the clip in a DVD-compliant format. After you import the intro clip into Adobe Encore DVD, create a timeline for the clip and set the timeline as DVD first play, as outlined in the following section.

> **TIP** *Adobe Encore DVD gives each new timeline you create the default name of "Untitled Timeline" appended by the next available number. You can give a timeline a unique name by selecting the timeline, right-clicking, and then choosing Rename from the shortcut menu to open the Rename Timeline dialog box. Enter the desired name for the timeline and then click OK.*

Set Timeline as DVD First Play

After you create a timeline, you can set the timeline as DVD first play. When you set a timeline as first play, the timeline plays as soon as the finished DVD is inserted and loads the presentation into a set top DVD player. To set a timeline as first play, follow these steps:

1. Choose Window | Project. Alternatively, you can choose Window | Timelines.

2. Select the timeline that you want to set as first play.

3. Choose File | Set As First Play. Adobe Encore DVD sets the timeline to first play. Alternatively, you can select the timeline, right-click, and then choose Set As First Play from the shortcut menu. First play timelines as designated

by an icon that looks like a round button with the universal Play icon, as shown next:

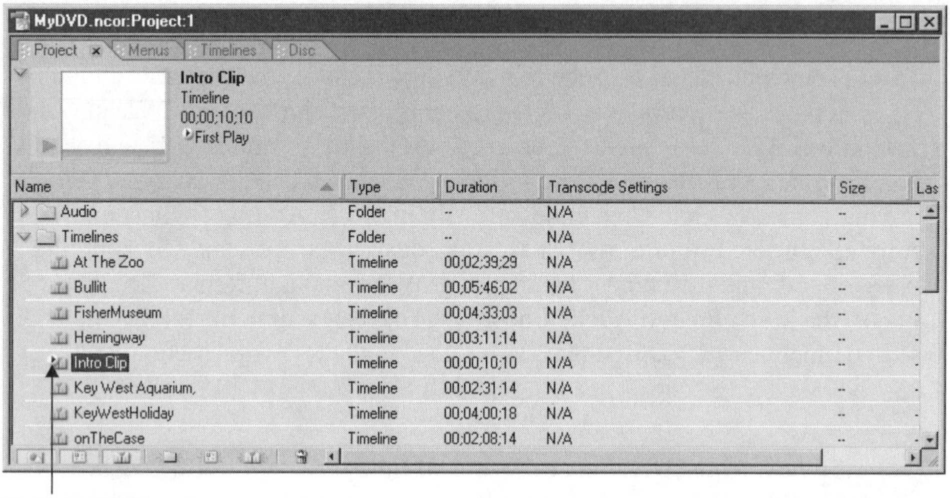

First play timeline

If after previewing your DVD, you decide a menu or different timeline would serve better as first play, select the current DVD First Play, choose Edit | Clear First Play, and then select the desired timeline or menu and set it as First Play.

Remember to set another timeline or menu as First Play before building the project. A project without a First Play will not play when inserted in a set top DVD player. If you know which timeline or menu you want to serve as the new First Play, select the timeline or menu in the Project tab, right click, and then choose Set As First Play from the shortcut menu. Setting a new First Play automatically clears the old one.

Explore the Timeline Window

You use the Timeline window to set chapter points, poster frames, and more. When you double-click a timeline, it is displayed in the Timeline window, shown next. You use the Timeline window tools to zoom in and out on the timeline, as well as navigate to a specific frame in the timeline. You can also navigate to specific frames using the Current Time Indicator.

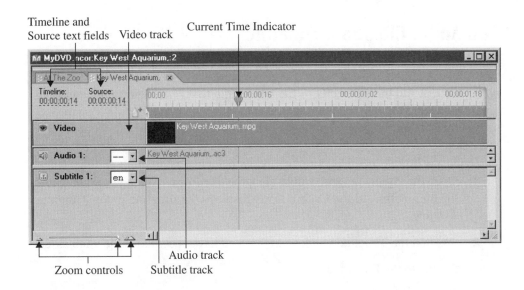

Timeline and Source text fields   Video track   Current Time Indicator

Zoom controls   Subtitle track   Audio track

Add Assets to a Blank Timeline

If you prefer, you can create blank timelines and then add project assets to the timeline. You can add one video asset or multiple image assets to a blank timeline. To populate a blank timeline, follow these steps:

1. Choose Window | Project to open the Project tab.

2. Double-click a blank timeline to display it in the timeline window.

3. Position the Timeline window so that it and the Project tab are visible.

4. Drag a video or image asset from the Project tab and drop it on the blank timeline. Adobe Encore DVD snaps the asset to the beginning of the timeline and displays the first frame of the timeline in the Monitor window.

NOTE   *If the blank timeline is being used for image assets, you can drag additional images to the timeline as needed. When you add additional images to a timeline, Adobe Encore DVD creates a new chapter point for each image you add. You can also select multiple images from the Project tab and drop them on a timeline. Images will be arranged in the order in which they are sorted in the Project tab. If you do not delete the default chapter points Adobe Encore DVD creates for each image, you are limited to 99 images per timeline. If you delete each chapter point except the first, you can put as many images on a timeline as needed until you exhaust the available DVD disc space.*

Add Audio Clips to a Timeline

Each timeline you create can handle up to eight audio tracks. If you use an application like Adobe Audition or Sony Sound Forge to create alternate audio tracks for video clips, you can add them directly to the applicable timeline. If your video editing application supports AC-3 encoding, you can render your audio and video tracks separately and marry them on a timeline. To add an audio track to a timeline, follow these steps:

1. Open the Project tab and double-click the timeline to which you want to add the audio clip. Adobe Encore DVD displays the timeline in the Timeline window.

2. Position the Timeline window so that you have access to Project tab assets.

3. Switch to the Project tab.

4. Drag the desired audio clip from the Project tab to the Timeline window.

5. Release the mouse button to add the audio clip to the timeline. If no audio track exists, Adobe Encore DVD creates a new one. If audio tracks exist on the timeline, Adobe Encore DVD creates an additional one, appending it with the next available track number. If you have already added eight audio tracks to the timeline, your cursor will be a circle with a diagonal slash indicating you will not be able to add additional tracks to the timeline.

Remove an Audio Track from a Timeline

You can remove an audio track from a timeline when it is no longer needed. To remove an audio track from a timeline, follow these steps:

1. Choose Window | Timeline to open the Timeline tab.

2. Double-click the timeline that contains the audio track you want to remove.

3. Click the audio track you want to remove.

4. Right-click and choose Remove Audio Track from the context menu. Alternatively, you can choose Timeline | Remove Audio Track.

Clear an Audio Track

You can also clear an audio track. When you clear an audio track, you remove the audio clip, yet the audio track remains as part of the timeline. After clearing an audio track, you can add a different audio clip to the track. To clear an audio track, follow these steps:

1. Choose Window | Timeline to open the Timeline tab.

2. Double-click the timeline that contains the audio track you want to clear.

3. Click the desired audio track, right click, and then choose Clear from the shortcut menu. Alternatively, you can press DELETE or the BACKSPACE key.

 When you select an audio track, you can also cut or copy the track's audio clip by selecting the command from the shortcut menu. After cutting or copying an audio clip, you can paste it into a blank audio track in any timeline by selecting the audio track and choosing Paste from the shortcut menu.

Navigate the Timeline

After creating a timeline for a video asset, you navigate to desired frames to set chapter points and poster frames. The Adobe Encore DVD design team has engineered a tremendous amount of versatility into the application. This trend continues when it comes to the task of navigating a timeline. The following sections discuss the different ways you can navigate a timeline and change timeline magnification.

Manually Navigate the Timeline

You can preview the frames in a timeline manually. You can manually navigate the timeline using one of these methods:

- Drag the Current Time Indicator along the timeline. This is also known as *scrubbing the timeline.* If you have the Monitor window displayed as you drag the timeline, the window updates to show you the frame the Current Time Indicator is currently over.

- Click the right arrow key to advance forward a frame at a time.

- Click the left arrow key to advance backward a frame at a time.

- Click the frame to which you want to advance. This advances the play head to the selected frame.

■ Press the SPACEBAR to play the timeline. Press the SPACEBAR again to pause
   the timeline. When you press the spacebar, you can preview the timeline
   in the Monitor window. The first audio track attached to the timeline plays
   as well.

Navigate to a Specific Frame

Across the top of the Timeline window are two text fields labeled Timeline and
Source. These fields display the current frame Timeline and Source timecode in
the following format: hours ; minutes ; seconds ; frames. The Source text field displays
the frame of the original clip as rendered from a video-editing application. Timeline
and Source display the same values unless you have trimmed a clip in Adobe Encore
DVD, in which case the Source text field displays the timecode from the original
track. To navigate to a specific frame, do the following: Click the Timeline or Source
text field. Adobe Encore DVD highlights the frame timecode and your cursor icon
changes to an I-beam.

1. Enter the timecode for the frame to which you want to advance. You can
   manually enter the entire timecode as hours ; minutes ; seconds ; frames, or
   you can click-and-drag to select the part of the timecode you want to change
   and then enter a different value.

2. Click ENTER. Adobe Encore DVD advances the Current Time Indicator to
   the frame that corresponds to the timecode you entered and updates the
   timecode in each field.

TIP   *Move your cursor toward the Timeline or Source text field. When your cursor
       icon changes to a hand and a dual-headed arrow, drag right to advance
       forward a frame at a time or left to go backward a frame at a time. Release
       the mouse button when you have reached the desired frame.*

Change Timeline Magnification

When you're working with a lengthy timeline, it takes considerable time to navigate
to the proper frame. Instead of navigating the timeline frame by frame, you can
speed up navigation by changing timeline magnification. You can zoom out to see
more of the timeline. After you've selected a frame that's near the one you want to
use as a chapter point or poster frame, you can zoom in on the timeline and navigate
to the desired frame using one of the techniques discussed previously. You change

timeline magnification by using the Timeline Zoom slider, as shown in the following illustration:

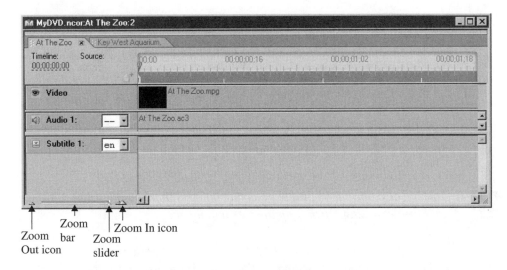

To change timeline magnification, do one of the following:

■ Click the Zoom Out icon to zoom out to a lower magnification level. Click as needed to zoom to lower magnification levels.

■ Click the Zoom In icon to change to the next highest level of magnification. Click as needed to zoom in tighter on the timeline.

■ Click the Zoom slider and drag left to zoom out, right to zoom in.

■ Click a point on the Zoom slider bar to zoom to that level of magnification.

NOTE
You can also zoom in and out on the timeline by selecting the Zoom tool from the toolbar. Click the timeline to zoom to the next highest level of magnification. Press the ALT key and click the timeline to zoom out.

Preview a Timeline

Another manner in which you can select frames for chapter points and poster frames is by previewing a timeline. When you preview a timeline, you can play it in its entirety or pause at a frame that might be a candidate for a chapter point or poster frame. You can then use one the methods listed in the "Navigate the Timeline" section to navigate frame-by-frame to choose the ideal frame for a chapter point or poster frame. To preview a timeline, follow these steps:

1. Choose Window I Timelines to open the Timelines tab.

2. Double-click the timeline you want to preview. Adobe Encore DVD opens the Timeline window and Monitor Window.

3. Click the Play button to preview the timeline.

You can also play a timeline by opening it, and then pressing the SPACEBAR. Press the SPACEBAR again to pause the timeline. Every time you press the SPACEBAR, you play or pause the timeline.

 *You can also open a timeline by selecting it in the Timelines tab, right-clicking, and then choosing Open from the shortcut menu.*

If you're working several timelines, each timeline you select appears as a different tab in the Timeline window, as shown in the following illustration. You select a timeline by clicking its tab. You can then preview the timeline in the Monitor window. If the Monitor window is not open, choose Window I Monitor.

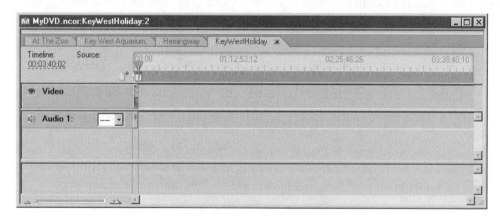

About the Monitor Window

You play a timeline by using the controls in the Monitor window. The Monitor window shown in the following illustration features VCR-like controls that you use to play, pause, advance forward, or skip backward along the timeline. Note that the controls to advance forward and skip backward are not available when you add an AVI file to a timeline.

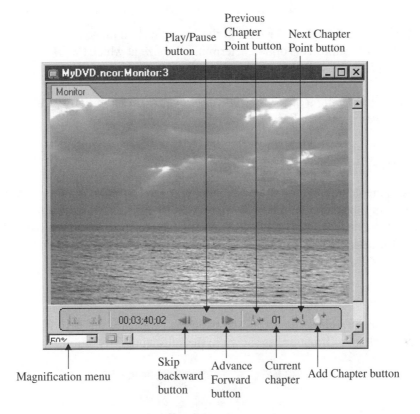

To navigate the timeline using the Monitor window, open a timeline in the Monitor window as described in the previous section, and then do one of the following:

■ Click the Play button to play the timeline. After you click the Play button, it becomes the Pause button.

■ Click the Pause button to pause the timeline. After you click the Pause button, it becomes the Play button.

■ Click the Advance Forward button to advance to the next GOP (Group Of Pictures) header.

■ Click the Skip Backward button to go to the previous GOP Header.

■ Click the Add Chapter button to add a chapter point at the current frame. For more information on chapter points, see the upcoming "Set Chapter Points" section.

■ Click the triangle to the right of the Magnification menu and choose a preset magnification. This option determines the size at which the Monitor window displays the video clip in the timeline.

 *After opening a timeline, you can play it by pressing the* SPACEBAR. *Repeat to pause the timeline.*

View Timeline Properties

After you create a timeline for a video clip, you need to perform other tasks such as specifying the end action for the timeline. Timeline and menu actions will be discussed in detail in Chapter 9. You can set the end action for a timeline and view other information about a timeline by viewing its properties. To view timeline properties, follow these steps:

1. Select the timeline whose properties you want to view.

2. Choose Window | Properties to open the Properties palette and display the timeline information, as shown in the following illustration:

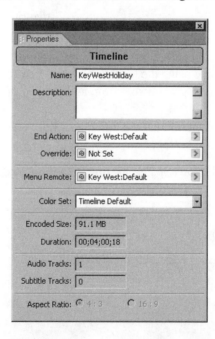

TIP *You can rename a selected timeline while viewing its properties by entering a new name in the Name field. If desired, you can enter a description in the Description field.*

Edit a Timeline

After you create a new timeline for a video asset, you can edit the timeline. When you edit a timeline, you can set chapter points, set poster frames, and trim clips. You can also delete video, audio, or image assets from a timeline. To edit a timeline, follow these steps:

1. Choose Window | Timeline to access the Timeline tab.

2. Double-click the desired timeline to display it in the Timeline window and Monitor window.

NOTE *If you are editing several timelines and the Timeline window is already open, you can select a timeline by clicking its tab.*

3. Edit the timeline by performing one or more of the tasks outlined in the following sections.

Set Chapter Points

When you create a timeline for a video asset, Adobe Encore DVD creates a chapter point of the first frame of the timeline. You can add additional chapter points as needed and then link menu buttons to the chapter points, as I describe in Chapter 9. Remember, you have a maximum of 99 chapter points per timeline and there must be at least 15 frames between adjacent chapter points. To set a chapter point, follow these steps:

1. Open the timeline for which you want to set chapter points using one of the methods outlined earlier in this chapter. Make sure you can view the Timeline window and the Monitor window.

2. Navigate to the frame where you want a new chapter to begin.

TIP *If the timeline is zoomed in to sufficient magnification, you can select an I-frame for a chapter point. I-frames are designated by white hash marks on the bottom of the time ruler. Note that I-frames are present only when an MPEG video is in the timeline. You can set a chapter point anywhere in an AVI video as long as the adjacent chapter points are at least 15 frames apart.*

3. Click the Add Chapter button. Adobe Encore DVD places a chapter point marker on the chapter point frame, as shown in the following illustration. Selected chapter points are highlighted in red; deselected chapter points are off-white.

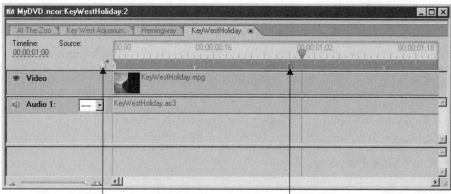

Add Chapter button   Chapter point marker

 *You can use the Monitor window to create chapter points. Use the Play, Advance Forward, and Skip Backward buttons to select the frame where you want the chapter to begin and then click the Add Chapter button.*

When you navigate the timeline to find the frame where you want a chapter to begin, you may find it helpful to play the entire timeline before you tackle the task of designating the frames where you want chapters to begin. Pause the timeline when you see a frame that's a likely candidate for a chapter point and note the timecode. Continue in this manner until you've got a rough idea of which frames will be chapter points. Then you can use the Timeline text field to enter the timecode for the first chapter point. You can then navigate frame by frame to fine-tune the process, or you can select a nearby I-frame for the chapter point.

Name Chapter Points

When you create a chapter point, it is designated by a number on the timeline. The default name for the new chapter point is *Chapter* appended by the next available chapter number for the timeline. After you designate chapter points, you have to link menu buttons to chapter points. If you don't change Adobe Encore DVD's default chapter numbering, you'll have a hard time figuring out what each chapter is about. For example, remembering what happens on Untitled Timeline 3, Chapter 4 can be a daunting task. You can quickly identify what happens at each chapter point if you give the chapter point a unique name. To name a chapter point, follow these steps:

1. Choose Window | Properties.

2. Create a chapter point using one of the previously outlined methods.

3. In the Properties palette, select the default chapter name.

4. Enter a new name for the chapter. This task will be easier if you arrange the Timeline window, Monitor window, and Properties palette so they are all readily accessible, as shown in Figure 6-1.

   Your task will be even easier if you name a chapter point right after you create it. After all, you've just previewed the timeline and viewed the frame you selected

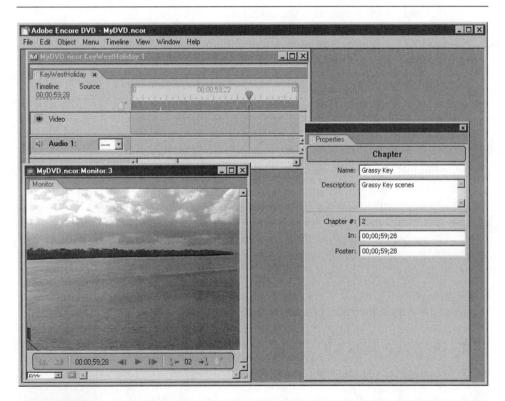

FIGURE 6-1   You can streamline your workflow if you name each chapter point.

for a chapter point, and the footage is fresh in your memory. Choose a unique name that matches the chapter footage, for example: Chase Scene.

If you don't name chapter points as you create them, you can rename any chapter point at a later date. Select the desired chapter point, open the Properties palette and enter the desired name in the Name: field.

Edit Chapter Points

When you author a DVD, you go through many steps before building the project. During the review phase, you may find that some chapter points are no longer needed or that you need to move a chapter point to a different frame.

To move a chapter point to a different frame:

1. Open the timeline whose chapter points you want to edit.

2. Select the chapter point you want to move. Remember, chapter points are designated by markers on the timeline.

3. Move the Current Time Indicator to the frame where you want to move the chapter point.

4. Drag the chapter point to the Current Time Indicator.

Alternatively, you can select a chapter point and drag it to the desired frame. However, when you first align the Current Time Indicator to the desired frame, you have two benefits. First you can see the timecode for the frame; second, you can view the frame in the Monitor window.

TIP   *You can navigate to a chapter point by choosing Window | Timelines to display the Timelines tab. Select a timeline and then double-click the chapter point in the bottom half of the Timelines tab. Adobe Encore DVD opens the timeline in the Timeline window with the chapter point frame selected. The image from the frame is also displayed in the Monitor window.*

To remove a chapter point from the timeline:

1. Open the timeline whose chapter points you want to edit.

2. Navigate to the chapter point you want to remove and select it.

3. Choose Edit | Clear. Alternatively, you can press DELETE.

Set Poster Frames

When you add a chapter point to a timeline, Adobe Encore DVD creates the chapter poster frame on the same frame. The poster frame is displayed as a thumbnail on a graphic menu button. However, sometimes the poster frame isn't indicative of what viewers see if they choose to play the chapter. As an example, if the chapter begins with animated text, the text appears on the poster frame and is anything but legible when the poster frame is viewed as a menu button thumbnail on a television screen. You can choose a different frame as the poster frame by following these steps:

1. Open the timeline whose chapter point poster frames you want to edit.

2. Navigate to the chapter point whose poster frame you want to set.

3. Navigate to the frame you want displayed as the poster frame.

4. Select the chapter point marker. When you select a chapter point marker, it becomes red in color.

5. Choose Timeline | Set Poster Frame. Alternatively, you can right-click and choose Set Poster Frame from the shortcut menu. When you set a chapter poster frame on a frame other than the chapter point frame, Adobe Encore DVD adds a rectangular poster frame marker to the timeline. The poster frame marker has the same number as the chapter point for which the poster frame is created, as shown in the following image:

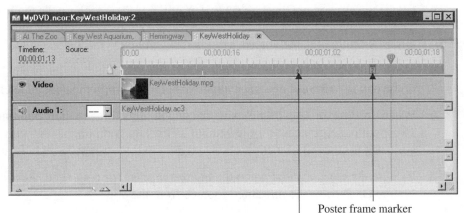

Poster frame marker
Chapter point marker

NOTE *If you're editing a long timeline, and the desired poster frame is a significant number of frames away from the chapter point frame, you may have to zoom out on the timeline to see both at once.*

You can set chapter points, delete chapter points, and set poster frames by selecting a frame, right-clicking, and then choosing the desired command from the shortcut menu.

Trim Video and Audio Clips

Your best bet is to use a video-editing application to trim unwanted portions of video clips and audio-editing software to trim unwanted portions of audio clips prior to importing the clips into Adobe Encore DVD as assets. You can, however, trim the beginning and ending point of any video or audio asset. You cannot split a clip.

You trim video and audio clips in the Timeline window. If you are editing a clip with both audio and video, you cannot edit each track independently, even though they do show up as separate tracks in the timeline window. When you trim the head of a clip, you change the In point, and when you trim the tail of a clip, you change the Out point. To trim a video or audio clip, follow these steps:

1. Open the timeline that contains the tracks you want to trim.

2. Drag the Current Time Indicator to the frame at which you want the video to begin or end. As you drag the Current Time Indicator, the image for the current frame is displayed in the Monitor Window.

3. Release the mouse button when the desired frame appears in the Monitor window.

4. Click the Selection or Direct Select tool. You'll find these tools on the toolbar. The Selection tool is the angled black arrow at the top of the toolbar and the Direct Select tool is the angled white arrow just below the Selection tool.

5. Move your cursor to the head of a track clip to trim footage from the beginning of the clip or to the tail of a track clip to trim footage from the end of the clip.

6. When your cursor icon becomes an arrow with a red bracket (left pointing arrow with a left bracket for the head of a clip, right pointing arrow with a bracket for the tail of a clip), click and drag to the Current Time Indicator.

7. Release the mouse button when the bracket reaches the play head. Adobe Encore DVD trims the unwanted footage from the clip.

NOTE *If, after trimming a clip, you decide you did not select the desired In or Out point, you can trim the clip again. Select either the Selection or Direct Select tool and move it toward the end of the clip you trimmed. When your clip becomes a bracket (left for the head of a clip, right for the tail of a clip) with a dual-headed arrow, you can drag right or left to further trim the clip, or restore some of the footage if you trimmed more than the desired number of frames from a clip.*

Edit Tracks and Clips

When you create a timeline for a video asset, you can add additional tracks, delete tracks, delete clips, and so on. You can edit a timeline as needed. You can perform the following tasks on an open timeline:

- Clear a selected video or audio clip from a track by choosing Edit | Clear. Alternatively, you can press DELETE. After doing either, Adobe Encore DVD clears the clip from the track, which you can then replace with a different clip.

- Remove a selected audio track by choosing Timeline | Remove Audio Track.

- Add a new audio track by choosing Timeline | Add Audio Track. This adds a blank audio track to the timeline, to which you can add an audio asset.

- Add an audio track by dragging an audio asset from the Project tab to an open timeline. This adds a new audio track and audio asset in one fell swoop.

- Copy an asset from the timeline to the clipboard by choosing Edit | Copy or by pressing CTRL-C.

- Paste a copied asset into a new timeline by choosing Edit | Paste or by pressing CTRL-V.

Edit Timeline Graphics

You can add as many still images as needed to a timeline. However, if you accept the default chapter point Adobe Encore DVD creates for each image you add to a timeline, you are limited to 99 images. As noted previously, you can delete a chapter point by selecting it, and then choosing Edit | Clear, or by pressing DELETE. Alternatively, you can delete a chapter point by right-clicking and then choosing Delete Chapter Point from the shortcut menu.

When you add additional graphics to a timeline, they snap to the end of the timeline by default. You can, however, change the position of images, modify the duration for which an image is displayed, copy and paste images, and so on.

Modify Display Duration

When you place images on a timeline, they display for six seconds by default when the timeline is played back. You can modify the default duration by choosing Preferences | Timelines and then entering a different value, as outlined in Chapter 3. You can also modify image display duration by doing the following:

1. Open the timeline that contains the images whose display duration you want to modify.

2. Click the Selection or Direct Select tool.

3. Move your cursor towards the tail of the image clip whose duration you want to modify.

4. When your cursor becomes a left pointing arrow with a red left bracket, click and drag to the right to increase the duration or left to decrease the duration.

If you want to increase the duration of a still image, there must be a blank space next to it on the timeline. If there are images adjacent to the image for which you want to increase the display duration, you can open a space by selecting all images that appear after it on the timeline and then dragging them to the right.

Modify Image Location on Timeline

When you display several images on a timeline, you can change the order in which they are displayed. To move an image to a different position in the timeline:

1. Open the timeline that contains the image(s) you want to move.

2. Click the Selection or Direct Select tool.

3. Click the image you want to move and drag it to a different position on the timeline.

When you move images along the timeline, you may experience gaps between images. If this occurs, move the images until they are against each other. You may find it helpful to remove the chapter points that are created when you add images to a timeline.

Copy and Paste Images

If you have still images that you want to display in other positions on the timeline, you can do so by copying and pasting. You can also copy an image and paste it into a new timeline or an existing timeline that contains still images. Remember you cannot mix video assets and still images on the same timeline. To copy and paste an image:

1. Select the timeline that contains the image you want to copy and paste.

2. Click the Selection or Direct Select tool.

3. Select the image you want to copy and then choose Edit | Copy. Alternatively, you can right-click and choose Copy from the shortcut menu or press CTRL-C.

4. Drag the Current Time Indicator to the timeline position where you want to paste the image. If you precisely align the Current Time Indicator to the end of an image, the image you paste will be perfectly aligned to the previous image. You may find it helpful to zoom in on the timeline to align the Current Time Indicator to the desired frame.

5. Choose Edit | Paste. Alternatively, you can right-click and choose Paste from the shortcut menu or press CTRL-V.

How to ...   Create a Slide Show

As you already know, you can create timelines that contain only still images. Still images are useful when you want to display object details. You can also create a timeline with many images and display them in succession, just like a slide show. To create a slide show, import your slide show images into the project as assets. Create a new timeline and drag the slide show images from the Project tab to the new timeline. Select the second chapter point, and then Shift-click all remaining chapter points except the first one. Choose Edit | Clear to remove the unnecessary chapter points. Link a menu button to the slide show timeline and your task is done.

Summary

In this chapter you learned to work with timelines. You learned to create timelines for video, audio, and still image assets. You learned to add chapter points to timelines, create poster frames, edit chapter points and poster frames, as well as edit the contents of timelines. In the next chapter, you'll learn to transcode timelines.

Chapter 7

Transcode Project Assets

How To...

- Transcode video assets
- Transcode audio assets
- Use preset transcode settings
- Create custom transcode settings

After you add audio and video assets to a project, you create timelines as outlined in Chapter 6. If your assets are all DVD compliant, you're ready to begin building navigation menus for your project. If they are not DVD compliant, you will have to transcode them. You may also have to transcode DVD compliant assets if there is not enough disc space to accommodate the material you want to present on the DVD disc.

You can let Adobe Encore DVD automatically transcode your assets when you build the DVD. However, if you do this, you run the risk of not having enough disc space for your project assets. If you use the Disc tab to analyze the disc space used by your project assets, you'll know whether you have enough space for the assets or whether you'll have to use different transcode settings to compress the files so they'll fit on the disc. In this chapter, you'll learn to transcode video and audio assets. You'll also learn to choose transcode settings for a project and create custom transcode settings for future projects.

What Is Transcoding?

As soon as you create a timeline for an asset, Adobe Encore DVD incorporates it as an asset so it will be in the final project build. If the asset has previously been rendered in a DVD-compliant file format, the asset does not need to be transcoded. If, however, an asset you add to a timeline is not DVD compliant, the asset will be transcoded when the project is built. When an asset is transcoded, it is converted to a DVD-compliant format. If the noncompliant asset is video, Adobe Encore DVD transcodes it as an MPEG-2 file. If the asset is audio, Adobe Encore DVD transcodes it using the default Dolby Digital audio encoding scheme. Other available audio encoding schemes are MPEG-1 Layer 2 or PCM. You can change the default audio encoding scheme to suit your projects by modifying Encoding preferences as outlined in Chapter 3.

Transcoding is similar to optimizing an image for an intended destination. For example, when you optimize a JPEG image for delivery on the Web, you compress the image to achieve a file size that will download quickly into the user's web browser.

When you transcode a video asset and then build a DVD, the video assets plays back at the specified bitrate. You can let Adobe Encore DVD take control of the transcoding noncompliant DVD assets, or you can choose your own transcode settings to suit your project or intended playback devices.

If you let Adobe Encore DVD take the reins, the application chooses the optimal bitrate that will fit all of the project assets on the disc size you specified when you set the project settings in the Disc tab. If your project contains only AVI video, the entire project will have to be transcoded. If you opt to set transcode settings, you may end up choosing a setting to achieve optimal video quality. You do this by choosing a transcode setting for an asset. You won't know the impact the new setting will have on your project until you actually transcode the asset. After transcoding, you can monitor the remaining disc space with the Disc tab.

If the new transcode settings negatively affect your project, you can always revert to the original file, and transcode the timeline with a different setting. However, transcoding lengthy timelines takes a considerable amount of time. In this regard, you're better off calculating the required data rate for the project as soon as you know the details about the amount of video and audio required for the project. To find out more about calculating video and audio data rate, refer to Chapter 4.

When to Transcode

You can transcode assets at any time; however, it makes good sense to transcode them after creating timelines. After you create timelines for your video assets, you'll be able to see how much space they'll use when transcoded for DVD. You do this by choosing Window | Disc. The disc space used and available is displayed in this tab, as shown in Figure 7-1. If the project assets exceed the available space on the media on which you're building the project, you can specify a lower quality transcode setting to reduce the file size of the transcoded assets.

You can also transcode assets that are DVD compliant. You should do this only when you're faced with a space problem. When you choose a lower quality setting to achieve a smaller file size, the quality of the media is degraded. However, if you create a custom setting that deviates only slightly from the automatic settings, you may be able to achieve the perfect compromise between quality and file size. Furthermore, if you're transcoding video assets for motion menus or an intro clip that was originally created in a vector illustration program such as Adobe Illustrator and then animated in Adobe After Affects, you may be able to get by with a lower quality setting, especially if the vector graphics don't contain elaborate fills. Remember, vector graphics are rasterized (converted to bitmap images) when you render them as video.

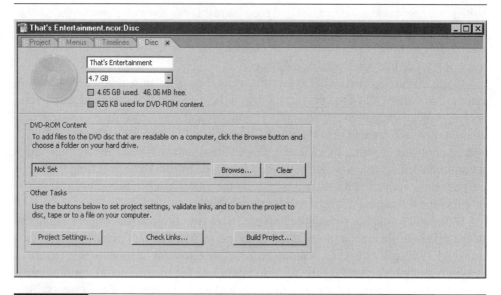

FIGURE 7-1   You can monitor available disc space using the Disc tab.

If you have images with millions of colors intermingled with your illustration, you won't achieve good results if you use the low quality settings.

Transcode Project Assets

When you transcode project assets, you can choose from a large variety of presets. You can let Adobe Encore DVD handle the transcoding automatically, choose a preset transcode setting, or create a custom transcode setting. When you transcode video assets, the available options are determined by the television broadcast standard you choose when you create the project. You can choose from CBR encoding or VBR encoding.

About CBR Encoding

CBR stands for *Constant Bit Rate*. When you choose CBR encoding, Adobe Encore DVD uses a fixed data rate for all video assets that are transcoded. CBR encoding is single pass, and therefore Adobe Encore DVD can transcode the asset quicker. However, you may end up with better quality video and will definitely have a smaller file size when you choose VBR encoding with the same target bitrate as a CBR setting.

About VBR Encoding

VBR stands for *Variable Bit Rate.* When you specify VBR encoding, Adobe Encore DVD makes one or two examining passes to determine the optimum bitrate for each video frame and then transcodes the asset. Video files are encoded using a data rate between a minimum and maximum bitrate. A target bitrate is also specified. When the video is encoded Adobe Encore DVD heeds to the target bitrate whenever possible. Video will be encoded below the target bitrate when applicable. The maximum bitrate gives you additional headroom should the video contain frames with more data than can be supported by the target bitrate. This is why you may end up with better video quality using VBR. The ability to use variable bitrates also almost always ensures a smaller file size.

About the Project Tab Transcode Settings Column

Each asset in a project is displayed in the Project tab. By default, the Transcode Settings column is visible, as shown in the following illustration. For each asset, one of the following transcode settings is displayed:

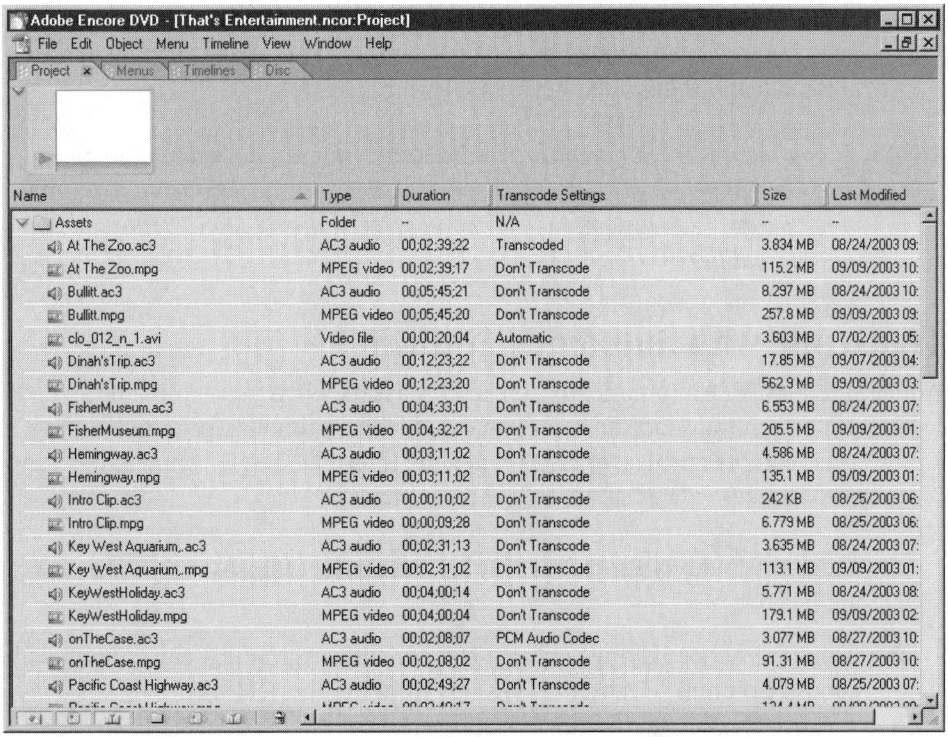

7

■ **Automatic**   This is the default setting for all non-DVD compliant assets. Adobe Encore DVD determines the optimum transcode setting for non-DVD compliant assets based on the number, length, and size of the project assets, while factoring in the available DVD disc space you specified when setting up the project. You can override automatic transcoding by choosing a transcode preset.

■ **Title of a transcode preset**   The title of the preset asset you select to override automatic settings is displayed in this column when you override automatic settings. If you do not override transcode settings, Automatic is displayed.

■ **Don't Transcode**   This information is displayed when you import DVD-compliant assets. If desired, you can use a preset transcode setting on DVD compliant assets. However, quality may suffer when you recompress the file. You will also receive a warning to this effect when transcoding DVD compliant assets.

■ **Transcoded**   This message is displayed when you have already transcoded an asset. If desired, you can revert to the original version of the file.

■ **N/A**   This message is displayed when an asset such as a still image or menu item cannot be transcoded by the DVD author. Still images and menus are automatically transcoded by Adobe Encore DVD.

> NOTE   *If you import DVD-compliant audio assets that do not match the audio encoding preference, Automatic appears in the Transcode column. If desired, you can override automatic transcoding by choosing Don't Transcode or choose a different transcode preset.*

Transcode with Automatic Settings

Adobe Encore DVD automatically transcodes assets when you build a project. However, you can transcode an asset you've added to a timeline prior to building a project. Doing this enables you to preview the transcoded asset while still editing the project. To transcode an asset, follow these steps:

1. Choose Window | Project to display the Project tab. All of your assets are listed.

2. If the Transcode Settings column is not visible, right-click any column and choose Columns | Transcode Settings from the shortcut menu.

3. Right-click the asset you want to transcode and choose Transcode Now from the shortcut menu.

Depending on the duration of the asset and the transcoding preset you choose, the transcoding process may take some time. Transcoding takes considerable processing power. In this regard, you are advised not to run other software programs while Adobe Encore DVD is transcoding a video asset. It's also important to note that you can't perform any other tasks in Adobe Encore DVD while transcoding.

Transcode with Preset Settings

After you create a new project, all the transcode settings for the specified television broadcast standard setting become available. You can override automatic transcode settings by doing the following:

1. Choose Window | Project to display the Project tab. All of your assets are listed.

2. If the Transcode Settings column is not visible, right-click any column and choose Columns | Transcode Settings from the shortcut menu.

3. Right-click the video asset, choose Transcode Settings and the transcode settings from the shortcut menu, shown here:

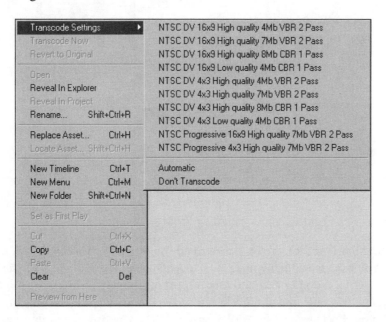

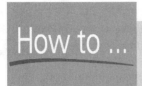 **Create a Copyright Warning**

If you author DVD discs that contain copyrighted material, you can copy protect a disc when you build the project, a task that is covered in Chapter 14. You can also create a warning that displays when the disc is first inserted in a DVD player. To create the warning, use an image-editing application such as Adobe Photoshop and create an image whose dimensions match the television broadcast standard for the disc you are creating. You can choose one of the Adobe Photoshop templates shown in the following illustration:

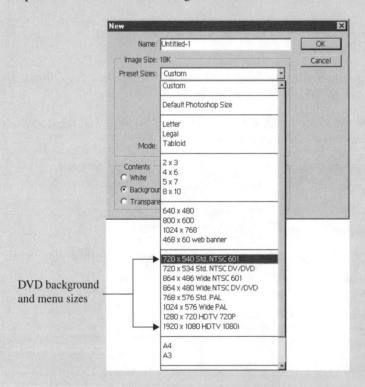

DVD background and menu sizes

Using the Text tool, create the text you want displayed when the DVD first plays. You can also add any images or logos you want displayed with the text. Remember to keep the text and any important images or logos within the 80 percent title safe area of a DVD disc. Save the image as a PSD file and then launch your DVD project in Adobe Encore DVD. Choose File | Import as Asset and select the copyright warning you created in Adobe Photoshop. Create a new timeline and add the copyright warning image to the timeline. Remember the default duration for a timeline image is six seconds. You can increase the duration by moving your cursor towards the tail of the image clip. When your cursor becomes a left pointing arrow with a red ending bracket, click, and drag the tail of the clip to the right to set the display duration of the image. Set the timeline for First Play, and the warning displays right after the disc loads.

Revert to Original File

When you transcode an asset, Adobe Encore DVD does not alter the original file. The transcoded version of the file is stored in the Project folder's Transcodes folder, which is a subfolder of the Sources folder. If desired, you can revert to the original file at any time by following these steps:

1. Open the Project tab. All transcoded assets are signified by the word *Transcoded* in the Transcode Settings column.

2. Right-click the transcoded asset and then choose Revert to Original from the shortcut menu.

CAUTION *When you build a DVD disc for an NTSC project and one or more timelines contains only MPEG sound, a warning dialog appears telling you that NTSC titles (timelines) must contain at least one Dolby Digital or PCM soundtrack. You can bypass the warning and build the disc; however, it may not be compatible with all DVD players.*

Modify Transcode Presets

If you're familiar with MPEG-2 compression, you can modify transcode settings to suit the projects you create. When you save a modified preset, Adobe Encore DVD adds *Copy of* before the name of the current preset. You can give the preset a unique name that reflects the type of projects or assets the modified preset will be used to transcode.

Understand Transcode Settings

Whether you let Adobe Encore DVD take the reins or you choose preset Transcode settings, the same video and audio options are being used to render the asset into a DVD-compliant format. The setting for each option is determined by the preset you choose. When you modify transcode presets to create custom presets, you work with the following settings:

- **Quality**   This setting determines the image quality of the video. Your options are from 1 (low quality) to 5 (high quality). Higher settings take longer to render.

- **Aspect Ratio**   This setting determines the video aspect ratio. Your options are either 4:3 (standard television screen) or 16:9 (widescreen format).

- **Frame Rate**   This setting is determined by the television broadcast standard of the preset you are modifying. If you are modifying an NTSC transcode setting, the frame rate is 29.97 fps while PAL transcode settings use a frame rate of 25 fps.

- **Program Sequence**   This setting determines the output scan mode. Your choices are Interlaced or Progressive. Most television sets use interlaced display. Interlaced display consists of a signal that has two fields for each frame of video. One field contains the information for the even lines of a video signal, while the other contains the information for the odd lines. The two fields are interlaced to form the picture. Progressive scan video processes all lines of a video signal at once, which results in a smoother and sharper picture. You should choose progressive only if the DVD disc will be played back using DVD players that support progressive scan and broadcast to a device that supports progressive display.

- **Field Order**   This setting is available if you choose interlaced for the program sequence. Your field order options are Upper or Lower.

- **Bitrate Encoding**   This setting specifies the compression technique used by the transcode setting. Your choices are CBR (Constant Bit Rate) or VBR (Variable Bit Rate).

- **Encoding Passes**   This setting is available when you choose VBR bitrate encoding. Your choices are 1 or 2. If you choose 1 pass encoding, Adobe Encore DVD will render the file quicker, but you may achieve a better video quality with a smaller file size by choosing 2 pass.

- **Bitrate**   This setting is available when you choose CBR for the Bitrate Encoding option. You can select a value between 3.0 MBPS and 9.0 MBPS.

- **Target Bitrate**   This setting is available when you choose VBR for the Bitrate Encoding option. This is the value that the encoder will attempt to use when transcoding assets. However, with VBR encoding you have leeway because the maximum bitrate gives you additional headroom when the encoder encounters complex video data.

- **Maximum Bitrate**   This setting determines the maximum bitrate the encoder can use. You can choose a value as high as 9.0 MBPS.

- **Minimum Bitrate**   This setting determines the minimum bitrate the encoder can use. You can select a value as low as 1.5 MBPS.

- **M Frames**   This setting determines the number of B frames between consecutive I- and P-frames. Your choices are: 2, 3, 4, 5, 6, 7, or 8.

- **N Frames**   This setting determines the number of frames between I-frames. You can select a choice from a drop-down list. The available choices are determined by the values you specify for M frames. The N Frame value is always an even multiple of the M frame value.

7

When you transcode an audio asset, the following options are in effect:

- **Codec**   This setting determines the compression codec Adobe Encore DVD uses to transcode audio assets. Your choices are Dolby Digital, Main Concept MPEG Audio, or PCM audio.

- **Audio Format**   This is not a setting but an informational display based on the audio codec you choose. If you choose Dolby Digital, the audio format is Dolby Digital Stereo, whereas Main Concept MPEG Audio uses the MPEG 1-Layer 2 audio format.

- **Bitrate**   This setting becomes available if you choose Dolby Digital or Main Concept MPEG Audio for the audio codec. If you choose Dolby digital, you can specify a bitrate between 128 kbps and 448 kbps. If you choose Main Concept MPEG Audio, you can specify a bitrate between 64 kbps and 384 kbps. Choose a higher bitrate setting for 5.1 Surround Sound files, for other files, choose the default setting for the encoder. If your DVD project audio consists of the spoken word, you may be able to get by with a lower setting.

- **Sample Rate**   This field displays the sample rate, which is always 48 kHz, regardless of the audio codec you choose.

NOTE   *If you choose PCM audio for audio encoding, you have no options. PCM audio also yields the largest file size, gobbling up valuable disc space; however, it does yield an audio track with excellent quality.*

Create a Custom Transcode Setting

Adobe Encore DVD provides a wide variety of transcode settings from which you can choose. However, the designers of the software also give you the option of modifying a preset and saving it as a custom setting. To create a custom Transcode setting, follow these steps:

1. Choose File | Transcode Settings | Modify Project Transcode Presets to open the Project Transcode Presets dialog box.

NOTE *Even though you are modifying a preset setting that you will use in other projects, you must have a project open in order for this menu command to be available.*

2. Click the triangle to the right of the Preset field and choose a preset to modify. After choosing a preset, Adobe Encore DVD displays a summary of the audio and video settings for the preset as shown next:

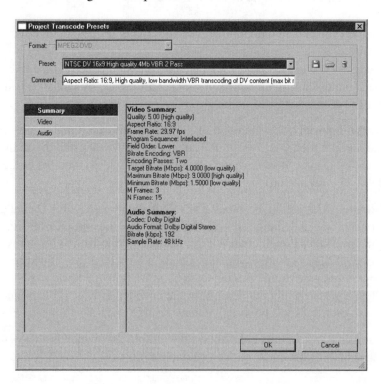

3. Click the Video tab to modify video settings for the preset. Adobe Encore DVD displays the current video settings for the preset, as shown next:

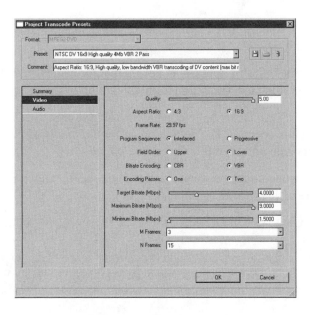

4. If desired, type a comment in the Comments field.

5. Specify video settings by adjusting each parameter. For more information on the available settings, refer to the "Understand Transcode Settings" section of this chapter.

6. Click the Audio tab to display the current audio settings for the preset. Adobe Encore DVD displays the audio settings currently in effect, as shown here:

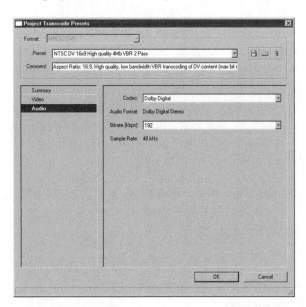

7. Click the triangle to the right of the Codec field and choose the desired audio encoding scheme.

8. Adjust the bitrate if applicable. For more information on audio settings, refer to the Understand Transcode Settings section of this chapter.

9. Click the Save Preset button that looks like a floppy disc. The Choose Name dialog box appears. The default name for the modified preset is the current preset name preceded by the words *Copy of.*

10. Type a name for the new preset and then press OK. Adobe Encore DVD saves the preset.

11. Click OK to exit the Project Transcode Presets dialog box.

After you exit the Project Transcode Presets dialog box, the new preset is available. To use the new preset, select an asset you want to transcode, choose File | Transcode | Transcode settings, and then choose the desired preset. Your custom presets appear at the top of the list.

Delete a Custom Transcode Setting

You can delete a custom transcode setting at any time. To delete a transcode setting that is no longer needed, follow these steps:

1. Choose File | Transcode Settings | Modify Project Transcode Presets to open the Project Transcode Presets dialog box.

2. Click the triangle to the right of the Preset field and then choose the preset you want to delete. The video and audio summary information for the preset is displayed. Before deleting the preset, review the summary information to verify that the preset is no longer needed.

3. Click the Delete Preset button that looks like a garbage can. Adobe Encore DVD displays a dialog box asking you to confirm deletion.

4. Click OK to delete the preset.

After deleting a preset, it no longer appears on the Project Transcode Settings list. If you find you've deleted the preset in error, you can undo the action by choosing Edit | Undo Edit Project Transcode Presets.

NOTE   *You cannot delete an Adobe Encore DVD system preset.*

 **Export and Import Transcode Presets**

If you create a custom setting that you want to share with other colleagues or archive for future use, you can easily do so. To export a transcode setting, follow these steps:

1. Choose File | Transcode Settings | Modify Project Transcode Presets to open the Project Transcode Presets dialog box.

2. Click the triangle to the right of the Preset field and select the preset you want to export. Adobe Encore DVD displays the summary information for the preset.

3. Press ALT while simultaneously clicking the Save Preset button. The Export Preset dialog box appears.

4. Enter a name for the preset and navigate to the file folder in which you want to store the exported preset.

5. Enter a name for the preset and click OK. Adobe Encore DVD exports the settings as a VPR (Video **PR**eset) file.

6. Click OK to exit the Project Transcode Presets dialog box.

When you import a VPR file, it is added to the Transcode Presets list as a custom setting. To import a VPR file, follow these steps:

1. Choose File | Transcode Settings | Modify Project Transcode Presets to open the Project Transcode Presets dialog box.

2. Click the Import Preset button that looks like a file folder.

3. Navigate to the directory in which the preset is stored, select the preset, and then click Open. The Choose Name dialog box appears.

4. Enter a name for the imported preset and click OK.

5. Click OK to exit the Project Transcode Presets dialog box.

After you import a VPR file, it is added to the Project Transcode list and is ready for immediate use.

Delete All Custom Transcode Settings

If desired, you can delete all of your custom transcode settings. However, before deleting all of your custom transcode settings, you may want to review each setting and export the ones you think may be beneficial in the future. To delete all custom transcode settings follow these steps:

1. Choose File | Transcode Settings | Modify Project Transcode Presets to open the Project Transcode Presets dialog box.

2. Press CTRL-ALT while simultaneously clicking the Delete Preset button. Adobe Encore DVD displays a dialog box asking you to confirm deleting all presets.

3. Click Yes to delete all presets.

4. Click OK to exit the Project Transcode Presets dialog box.

5. After performing the above steps, no custom presets appear on the Transcode Settings list. If you find you've deleted the presets in error, choose Edit | Undo Edit Project Transcode Presets.

Summary

In this chapter, you learned how Adobe Encore DVD handles transcoding. You learned how to transcode assets that are not DVD compliant and how to specify which setting is used to transcode an asset. You also learned to modify preset transcode settings to suit the DVD projects you create. In the next chapter, you'll learn to create menus for your DVD projects.

Part III

Work with Menus and Buttons

Chapter 8

Create and Edit Menus

How To...

- Create DVD menus
- Import assets as menus
- Create custom menus
- Work with text
- Edit menus

After you create a new project, import assets, and create timelines for your assets, your next step is to create navigation for your DVD project. When you create a DVD project with several videos, you design menus that enable viewers to navigate to various parts of your production. You create a main menu that lists the main parts of the DVD. These link to submenus that are comprised of buttons that link to various features, individual timelines, or chapter points of your DVD project.

Although the process of creating menus and submenus is virtually identical, this chapter will be devoted to creating menus. In this chapter, you'll learn how to create main menus for your DVD projects. You'll also learn to create text objects for your menus. Another topic of discussion is the powerful integration between Adobe Encore DVD and Adobe Photoshop.

About DVD Menus

As you probably know, DVD menus are navigation devices that the viewers of your DVD projects use to navigate to and view the various assets of your project. From within Adobe Encore DVD you can create menus using presets from the Library. Or, with a bit of forethought and the creative use of Adobe Photoshop, you can create stunning menus for your DVD project. Adobe Encore DVD also supports motion menus and animated buttons, which will be discussed in detail in Chapter 10. In Chapter 11, you'll learn to create custom DVD menus and buttons in Adobe Photoshop. In Adobe Encore DVD, all menus are Adobe Photoshop PSD files, which means you can quickly edit them in Adobe Photoshop.

After you create the main menu, you flesh out your project with submenus. As a rule, submenus contain buttons that link to timelines, or chapter points within timelines.

Create the Main Menu

Your main menu serves as the jumping-off point for your DVD project. From here, viewers will select the submenus that contain buttons that link to the various timelines and chapter points you've created. You can create a main menu by importing a Photoshop PSD file as a menu or by using a menu from the Library.

A main menu is like the table of contents in a book. When you create a main menu for a DVD project, you can display the table of contents over a colorful background that you create in an image editing application like Adobe Photoshop. Figure 8-1 shows a typical DVD main menu. This menu was created using a preset from the Adobe Encore DVD Library. The still image was imported as a project asset and added to the menu.

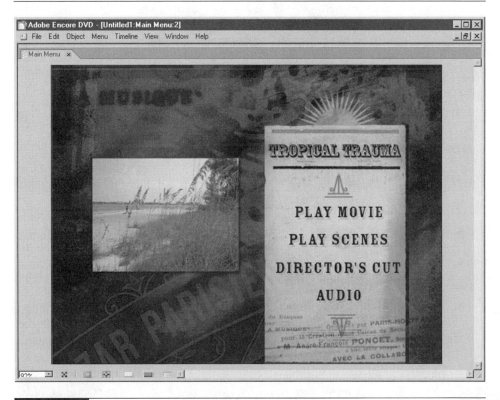

FIGURE 8-1   A main menu contains links to submenus.

Use Library Menu Assets

If you need to quickly create a professional looking menu for a DVD project, you can use a preset from the Adobe Encore DVD Library. To use a Library menu for your DVD project, follow these steps:

1. Choose Window | Menus. Adobe Encore DVD displays the Menus tab.

2. Choose Window | Library. The Library palette shown next opens.

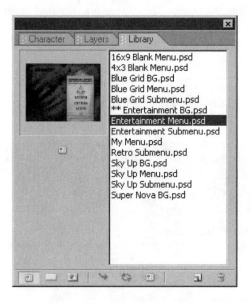

3. Click the Show Buttons and Show Images buttons to hide these Library items.

4. Click a menu item to preview it. A thumbnail version of the menu appears in the Library preview window.

5. Drag the selected menu item from the Library to the Menus tab. Alternatively, after selecting the desired menu, you can click the New Menu button at the bottom of the Library palette. The menu is added to the project and appears in the Menu Editor window.

TIP   *You'll find more menu presets, backgrounds, and buttons in the Goodies folder of Adobe Encore DVD installation disk. You can add a preset to the Library by opening the Library and then clicking the Add Item button. Navigate to the Goodies folder, select the item to add and then click Open.*

Create a New Menu

If you prefer, you can add a menu to your project by using a menu command. When you add a menu using a menu command, the default menu for your project's television broadcast standard is added to your project. To create a new menu using the Adobe Encore DVD Library default menu, choose Menu | New Menu. Alternatively you can press CTRL-M. After using either method, Adobe Encore DVD adds a new menu to your project and opens it in the Menu Editor.

Change the Default Menu

After you gain some experience with Adobe Encore DVD, create your own menus, or modify preset menus with Adobe Photoshop, you may find yourself using the same menu or a variation thereof for each project you create. If so, you can change the default menu item to your favorite menu by doing the following:

1. Choose Window | Library to open the Library palette. The current default menu is indicated by two asterisks (**) before its name.

2. Select the menu that you want to be the new default. A thumbnail version of the selected menu appears in the Library palette thumbnail preview window.

3. Right-click and choose Set as Default Menu from the shortcut menu. The selected menu becomes the default for creating new menus and two asterisks (**) precede the menu's name.

When you specify a new default menu, the change is applied immediately. When you next choose Menu | Create New Menu, your new default menu opens in the Menu Editor and will be added to the Menus tab as a project asset.

Use the Menus Tab

After you begin fleshing out your project by adding navigation menus, you use the Menus tab to select, edit, and preview menus. Whether you create a menu from scratch, use an Adobe Encore DVD Library preset, or create a custom menu in Adobe Photoshop and import it as an asset, the menu appears in the Menus tab, as shown in Figure 8-2.

After you begin adding menus to your project, you use the Menus tab to access the menus for editing and viewing in the Menu Editor window. At the top half of the Menus tabs is a list of the menus in your project. When you select an individual menu,

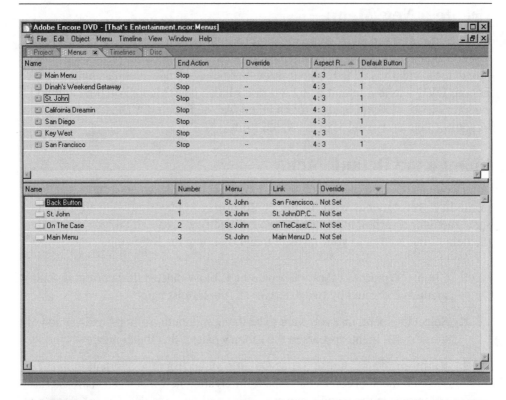

FIGURE 8-2 Your menu assets are neatly organized in the Menus tab.

the bottom half of the tab contains a list of all buttons used in the menu. You can display a menu in the Menu Editor window by opening the Menus tab and then doing one of the following:

- Double-click a menu name.
- Double-click a button, which displays the menu in the Menu Editor window, and the button is selected.

Use the Menu Editor Window

You use the Menu Editor window to add elements to a menu, add text to a menu, align elements in a menu, and so on. Adobe Encore DVD opens the Menu Editor when you create a new menu, import a menu, or double-click an existing menu in the Menus tab.

The Menu Editor window displays the menu you are currently editing. Each menu previously displayed in the Menu editor is designated by a tab with the menu name as shown in Figure 8-3. Click the tab to edit the menu.

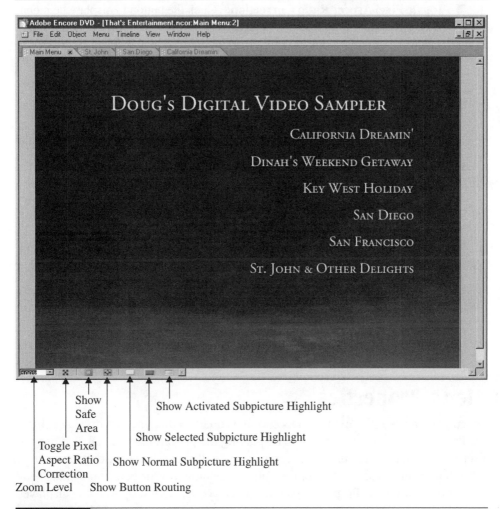

FIGURE 8-3   You use the Menu Editor window to edit project menus.

At the bottom of the Menu Editor window are buttons that you use to change your view of a selected menu as follows:

- **Zoom Level**   Controls the magnification of the menu. Click the triangle to the right of the window and choose a magnification from the drop-down list. Alternatively, you can type a value in the field.

- **Toggle Pixel Aspect Ratio Correction**   Click the button to display the menu, as it will appear on a television screen. Clicking this button produces no noticeable difference unless you're working with a menu that has the following dimensions: 704×480 pixels, 704×486 pixels, 720×480 pixels, or 720×486 pixels for NTSC menus; or 704×576 pixels, 720×576 pixels for PAL menus.

- **Show Safe Area**   Click this button to display action safe and title safe areas. When creating menus, keep all text elements within the innermost rectangle and strive to keep all important graphic elements within the outer rectangle.

- **Show Button Routing**   Click this button to display button numbers and button routing. Adobe Encore DVD automatically routes buttons; however, you can change routing to suit your project. Refer to Chapter 9 for more information on button routing.

- **Show Normal Subpicture Highlight**   Click this button to show the unselected subpicture highlight for all menu buttons.

- **Show Selected Subpicture Highlight**   Click this button to show the selected subpicture highlight for all menu buttons.

- **Show Activated Subpicture Highlight**   Click this button to show the activated subpicture highlight for all menu buttons.

Set Menu Properties

When you create a new menu, it has certain properties that you can display using the Properties palette. In addition to displaying the properties, you can also use the Properties palette to set menu properties (such as its name), the action that occurs when the menu stops, or whether the buttons are animated or not. You can also add a description using the Properties palette. When you add a description, it is displayed in the Menus tab Description column. For more information on describing an asset, refer to Chapter 3. In Chapter 10 you'll learn how to use the Properties palette to specify which video and audio track plays when you create a motion menu as well as how to use the palette to create animated buttons. The following illustration shows the Properties palette, as it appears when a menu is selected:

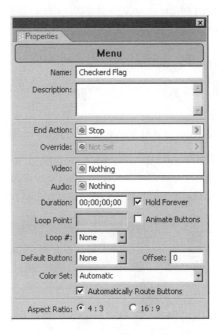

Name the Menu

When you create a new menu using a menu command or by choosing a menu item from the Library, it is displayed in the Menus tab by its default name. While the name is quite descriptive and appropriate when working in the Library, it can become quite confusing when you're working on a project with several menus. You can streamline your workflow considerably if you rename each menu to reflect the assets linked to the menu. To rename a menu, follow these steps:

1. Choose Window | Menus to display the Menus tab if it is not already open. Alternatively, if you're working in another tab, you can click the Menus tab to bring it to the front of the Project window.

2. Select the menu item you want to rename, right-click, and choose Rename from the shortcut menu. The Rename Menu dialog box appears.

3. Enter a name for the menu and then click OK. The new menu name now appears in the Menus tab. If any other project items link to the renamed menu, Adobe Encore DVD updates the links to reflect the new name.

You can also rename a menu from the Properties palette. This option is useful if you're working on several project items and have the Properties palette displayed. Remember the Properties palette displays the properties for the currently selected item. To rename a menu using the Properties palette:

1. Open the Menus tab as outlined previously.

2. Select the menu item you want to rename.

3. Choose Window | Properties to open the Properties palette. Alternatively, you can click the palette if it's already floating in the workspace.

4. Enter a new name for the menu in the Name field, as shown in the following illustration, and then press ENTER. The menu's new name appears in the Menus tab and all applicable project links are updated to reflect the new name.

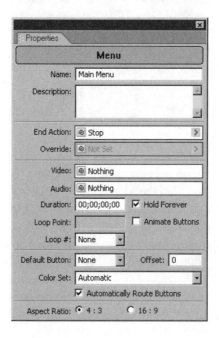

Set Menu Display Duration

By default, a menu displays forever or until the user activates a menu button. However, if you have a static menu, you can set the display duration for the menu. If a user doesn't interact with a menu button before the display duration elapses, the menu end action executes. You also set display duration when you create motion menus, a task that is covered in Chapter 10. To set menu display duration, follow these steps:

1. Open the Menus tab as outlined previously and select the menu for which you want to set display duration. Alternatively, you can select the menu from the Project tab.

2. Choose Window | Properties to display the Properties palette.

3. Click the Hold Forever checkbox to disable this option.

4. Enter the duration for which you want the menu to display in the Duration field. Remember the timecode is formatted as Hours;Minutes;Seconds;Frames. To display a menu for 10 seconds, you enter the following timecode: 00;00;10;00.

After you change the display duration for a static menu, you must modify the end action; otherwise, the screen will go black if the user does not activate a button before the duration time elapses.

NOTE *If you're creating a DVD that will run continuously and will be viewed by multiple viewers, such as a kiosk in a department store, set the default duration for the desired time, then set the end action to link to the main menu. That way the presentation will be ready for the next viewer.*

Set Menu End Action

When you modify the display duration of a static menu, you must change the End action from the default Stop action. You can change the End action so that another timeline plays or a different menu is displayed when the display duration elapses. To set the menu end action follow these steps:

1. Open the Menus tab and select the menu whose end action you want to modify.

2. Choose Window Properties to open the Properties palette.

3. Click the triangle to the right of the End action field and choose a link from the drop-down list, as shown next:

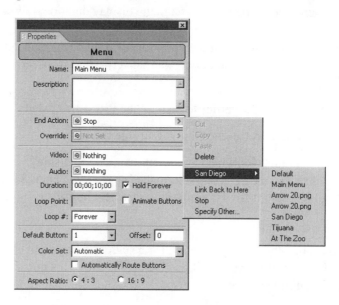

4. Click Specify Other if the desired link is not displayed. This opens the Specify Link dialog box shown next. Within this dialog box you'll find a listing for every menu, button, timeline, and chapter point in your project.

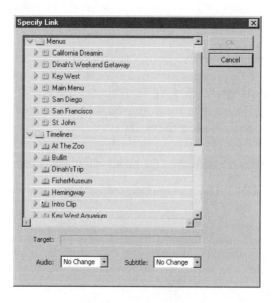

5. Click the desired link and then click OK to apply the new End action.

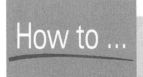

Create a Custom DVD Screen Saver

Most set top DVD players have a built in screen saver that functions similar to the screen saver on your computer. If a user does not interact with the DVD for a certain period of time, the DVD player launches an animation. You can create a custom screen saver by modifying the duration of menus in your project and specifying that the end action for each menu links to an animation timeline. The first step is to create a short animation. You can create the animation from video footage, or you can use an application such Adobe After Effects 6.0 to create a 15 or 20 second video clip. If you add a soundtrack to the animation, it can serve double-duty as entertainment and as a wake-up call for a viewer who may have dozed off while watching your DVD in the wee hours of the morning. After you create the animation, render it as a DVD-compliant video. In Adobe Encore DVD, import the video and create a timeline for the animation clip. Your next step is to modify the duration of each menu in your project and set the End action so that it links to your animation timeline. Modify the end action for your screen saver animation timeline to link to the main menu and you're done.

8

Adobe Encore DVD displays the new End action in the Properties palette and in the End Action column of the Menus tab. You also set links for End actions of timelines. The designers of Adobe Encore DVD have given you several ways to set links, which will be explored in detail in Chapter 9.

Set Menu as DVD First Play

Many DVD authors use a short animation as DVD first play to display information about their company or perhaps a copyright warning. If you prefer, you can set a menu as DVD first play. To set a menu as first play, follow these steps:

1. Click the Menus tab. Alternatively, you can press F10.

2. Select the menu that you want to set as DVD first play.

3. Right-click and then choose Set as First Play from the shortcut menu.

NOTE   *You can also set DVD first play by selecting the desired menu from the Project tab and then choosing Set as First Play from the shortcut menu.*

Import a PSD File as a Menu

If you own Adobe Photoshop, you can take advantage of the tight integration between Adobe Encore DVD and Adobe Photoshop to create custom menus for your DVD projects. When you create a menu in Adobe Photoshop, you use layers to create button sets complete with subpictures, as outlined in Chapter 11. When you save your custom menu as a PSD file, you can import the file into Adobe Encore DVD with all the layers and button sets intact and editable. To import an Adobe PSD file into a project as a menu asset, follow these steps:

1. Open the Project tab.

2. Choose File | Import as Menu Asset to open the Import as Menu dialog box. Alternatively, you can right-click anywhere inside the Project tab and choose Import as Menu from the shortcut menu.

3. Navigate to the folder in which you've stored the PSD file you created for use as a menu.

4. Click Open. Adobe Encore DVD imports the file and adds it to the Project tab and opens it in the Menu editor.

Edit Menus in Adobe Photoshop

When you flesh out a menu in Adobe Encore DVD, you add objects to the menu such as .png files from the Library, text objects, or images that you import into a project. Whether you begin with a blank menu, a library menu preset, or import a PSD file as a menu asset, Adobe Encore DVD stores the menu as a PSD *(Photoshop document)* file in the project folder. As the menu begins to take shape, you may find that you need to perform an edit that is not possible in Adobe Encore DVD, such as adjusting the hue and saturation of an image. You can edit a menu in Adobe Photoshop by following these steps:

1. Open the Project tab or the Menus tab.

2. Double-click a menu to view it in the menu editor.

3. Choose Menu | Edit in Photoshop. Alternatively, you can press CTRL-SHIFT-M. Adobe Photoshop launches and opens your menu, as shown in Figure 8-4.

4. Use the Adobe toolset and menu commands to perform the needed edits.

5. Choose File | Save. Adobe Photoshop saves the edited menu to the project folder. At this point, you can exit Adobe Photoshop to free up system resources.

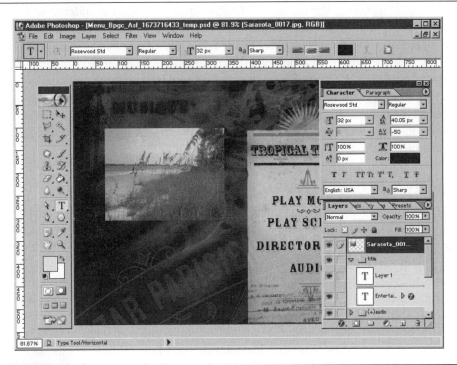

FIGURE 8-4   You can edit a DVD menu in Adobe Photoshop.

When you return to Adobe Encore DVD, your menu reflects the edits you performed in Adobe Photoshop. If you originally imported a PSD file for use as a menu, Adobe Encore DVD creates a copy of the file in the project cache folder. Therefore, when you edit the menu, all edits are performed on the copy and the original file is unaltered.

Edit Menus in Adobe Encore DVD

Whether you choose a preset menu from the Adobe Encore DVD Library or import a PSD file for use as a menu, you can edit certain facets of the menu directly in Adobe Encore DVD. You can add text to a menu and rearrange the order of layers in your menu, as well as move and resize objects.

Edit Layers

When you add text objects and buttons to your project, each object appears on its own layer. The layers are stacked in the order in which you add objects to the menu. In other words, when you create a new text object or add a button or graphic asset to a menu, it appears at the top of the stack, eclipsing all objects underneath it on lower layers. If you're familiar with image-editing programs such as Adobe Photoshop, you're familiar with the layer stacking order. When you edit a menu in Adobe Encore DVD, you work with objects on layers. You can edit objects directly in the menu using the Selection or Direct Select tool. When you select objects with either tool, you can resize and move them, tasks you'll learn in Chapter 9. You can also select an object by clicking the proper layer in the Layers palette. After selecting an object, button set, or layer set, you can arrange its order in the layer stack by using menu commands. You'll learn to arrange objects in Chapter 9.

Modify a Preset Menu

Adobe Encore DVD ships with an impressive array of preset menus and submenus that you find in the Library. These professionally-created menus are well suited for most projects. However, you can put your own stamp of originality on an existing menu by modifying it. You can modify a menu in Adobe Encore DVD by adding buttons to a menu, duplicating buttons, changing button text font face, style, and so on. You can also modify a menu by deleting a background and substituting it with a background image you imported into a project.

You can use Adobe Photoshop to modify a Library menu in your project. Simply select the menu you want to modify and then choose Menu | Edit in Photoshop. After you have a menu in Photoshop, you can use filters to modify the background image and use the Free Transform command to scale and rotate a menu button or object. Another possibility is using one of the image adjustment commands to modify the background image. For example, you can use the Hue and Saturation command to saturate, desaturate, or change the hue and give the background image a totally different look.

You could also modify the image by tinting it. Create a blank layer and then use the Paint Bucket tool to add a color to the layer. After adding color to the layer, you won't be able to see the background until you modify the opacity of the fill layer. Then it's a matter of choosing a blend mode, and voila, you've got a different looking background for your DVD menu.

If you know Adobe Photoshop well, you'll be able to quickly modify a menu background to suit your needs. If you're not a seasoned Adobe Photoshop veteran, experiment with the toolset and you'll find new ways to modify menus to suit your needs. You may also consider investing in one of these Osborne titles to hone up on your Adobe Photoshop skills:

Title	Author	ISBN
How to Do Everything with Photoshop(R) 7	Laurie McCanna	0072228342
Photoshop(R) 7: The Complete Reference	Laurie Ulrich	0072228687

8

> TIP   *You can create a menu from scratch in Adobe Photoshop. When you create a menu in Adobe Photoshop, you can use layer naming protocol that is recognized by Adobe Encore DVD to create buttons with different highlight states. Creating custom menus and button sets in Adobe Photoshop will be covered in detail in Chapter 11.*

Use a Blank Menu

If you're creating a DVD for a client and the client has supplied artwork and logos for the DVD, choose a blank menu, which is the default Library menu, unless you change the default as outlined in the "Change the Default Menu" section of this chapter. You can flesh out a blank menu with imported assets or Library items such as buttons and background graphics. Figure 8-5 shows an example of a menu that began life as a blank menu. The background image was imported as a menu asset. The title text and menu buttons were created with the text tool.

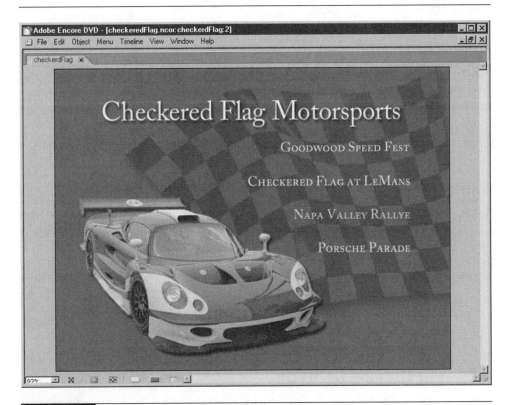

FIGURE 8-5   You can use a blank menu as the basis for a custom menu.

Duplicate a Menu

After you create a basic menu with a background image and title text, you can use this as a template for the other menus in your project. Duplicate this template to create enough menus for your project. You can add buttons to each duplicate menu as needed to suit the number of timelines or chapter points that will be linked to each menu. To duplicate a menu, follow these steps:

1. Open the Menus or Project tab.

2. Select the menu you want to duplicate.

3. Choose Edit | Duplicate. Alternatively, you can press CTRL-D.

When you duplicate a menu, Adobe Encore DVD creates a carbon copy of the menu and appends the filename with the word *copy*. You can assign a unique name to the menu using the Properties palette, or by right-clicking and then choosing Rename from the shortcut menu.

Save a Custom Menu

After you create a custom menu, you can save it to the Library for future projects. You save a custom menu to the Library in the same manner as you add an image to the Library. To save a custom menu to the Library, follow these steps:

1. Open the Project or Menus tab.

2. Choose Window | Library. Arrange the tab and palette so that both are accessible.

3. From the Project or Menus tab, select the menu that you want to save to the Library. You may want to rename the menu prior to adding it to the Library. Remember, you cannot rename a Library item.

4. Drag the selected menu from the tab into the Library. As you move your cursor into the Library, the cursor icon changes to a document with a plus sign (+).

5. Release the mouse button. Adobe Encore DVD adds the menu to the Library for future use.

Add Text Objects to Menus

Whether you start from scratch with a blank menu, import a PSD file as a menu, or work with a library preset, you will either be adding text objects to a menu or editing text objects. When you work with text in Adobe Encore DVD, you have the powerful

Adobe text engine as your ally, which enables you to have precise control over font size, font face, font color, and other text attributes such as kerning, vertical scale, horizontal scale, baseline shift, and so on. In the upcoming sections, you'll learn to create and edit text objects for your DVD menus.

Use the Text Tools

You use the text tools to create and edit text. You use the tool to create text and then define the text attributes using the Character palette. You have two text tools at your disposal: the Vertical Text Tool, and the Text Tool. To add text to a DVD menu, follow these steps:

1. Open the Menus tab or Project tab and then double-click the menu to which you want to add text. Adobe Encore DVD opens the menu in the Menus editor.

2. Select either the Text Tool or Vertical Text Tool, as shown next. When you select the Text Tool, your cursor becomes an I-beam. When you select the Vertical Text Tool, the I-beam is rotated 90 degrees clockwise, indicating that the text will flow from top to bottom.

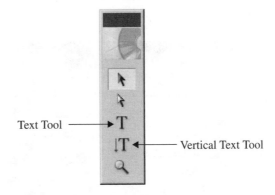

Text Tool ──────▶ T

│T ◀────────── Vertical Text Tool

3. Click the spot where you want the text to appear inside the menu.

4. Type the text. If needed, press ENTER to wrap text to a new line.

NOTE *When you add text to a menu, be sure to place it inside the title-safe area. You can view the title-safe boundaries by clicking the Show Safe Area button at the bottom of the Menu Editor. Remember, the title-safe area is designated by the inner rectangle.*

8

How to ... Create a Text Box

You can constrain the text to a rectangular area of your DVD menu by selecting either text tool, clicking the point where you want the text to begin, and then dragging diagonally inside the menu. As you move your cursor, Adobe Encore DVD displays a bounding box, that shows the current size of the text box. Release the mouse button when the text box is the desired size. When you enter text and reach the end of the bounding box, the text wraps to the next line. This option is convenient when you need to constrain text to a certain part of your menu.

Use the Character Palette

When you create a text object for a DVD menu, you use the Character palette to specify the font size, font face, font style, baseline shift, kerning, and much more. You can specify text attributes prior to entering text or edit the characteristics of existing text. From within the Character palette (shown next) you can specify the following text attributes for horizontal text:

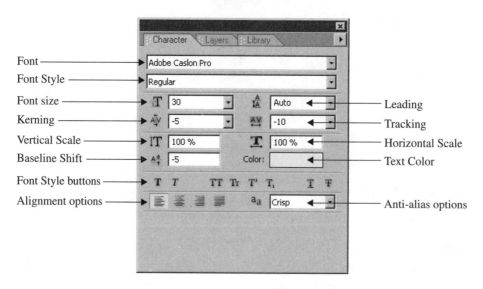

■ **Font**   Click the triangle to the right of the field to display a list of available fonts. Click a font to select it.

■ **Font Style**   Click the triangle to the right of this field to reveal a list of font styles, such as Bold or Italic. The available styles depend on the font you select. Click a style to select it.

> NOTE   *Font styles change in accordance with the font you have selected. Bold or italic may not be available for some fonts. If so, you can apply a style to text using one of the Font Style buttons in the Character palette.*

■ **Font Size**   Click the triangle to the right of this field to display a drop-down list of font sizes. Click a size to select it. Alternatively, you can manually enter the desired size.

■ **Leading**   Click the triangle to the right of the field and select a value. Leading sets the amount of space between lines of type. When you choose the Auto setting, Adobe Encore DVD sets the leading at 120 percent of the text font size.

■ **Kerning**   Determines the spacing between two letters. Place your cursor between the two letters you want to kern, and then click the triangle to the right of the field to choose a setting from the drop-down list. Choose a negative value to tighten up the spacing or a positive value to spread the letters apart.

■ **Tracking**   Click the triangle to the right of the field and choose a value from the drop-down list. Tracking modifies the spacing between characters of the selected text. Choose a positive value to increase spacing or a negative value to decrease spacing.

■ **Vertical Scale**   Enter a value greater than 100 to increase the vertical scale; enter less than 100 to decrease the vertical scale of the text. This value is a percentage of the original size of the text relative to the baseline.

■ **Horizontal Scale**   Enter a value greater than 100 to increase the horizontal scale; enter less than 100 to decrease the horizontal scale of the text. This value is a percentage of the original size of the text relative to the baseline.

■ **Baseline Shift**   Enter a positive value to raise selected text above the baseline; enter a negative value to shift text below the baseline. This value is in pixels. You can select several characters, a word, or several words from a line of text, and then enter a value for baseline shift to raise or lower the text relative to the other characters in the line of text.

- **Text Color**   Click the swatch to reveal the Color Picker. Drag inside the Color Picker to select text color. Alternatively, you can enter values in the H, S, B, or R, G, B fields to match a known color. Click the NTSC Colors Only checkbox to display only NTSC legal colors in the color picker.

- **Font Style buttons**   You can use the font style buttons to apply multiple styles to selected text. You can apply the following styles to text:

  - **Faux Bold**   Choose this style to simulate bold-faced text. This style is useful when the selected font does not include Bold as a style option. If the font includes a bold style, it is be available from the Style drop-down menu.

  - **Faux Italic**   Choose this style to simulate italicized text. This style is useful when the selected font does not include Italic as a style option. If the font includes an italic style, it is available from the Style drop-down menu.

  - **All Caps**   Choose this style to capitalize all letters of selected text regardless of the case selected when the letters were typed.

  - **Small Caps**   Choose this style to convert lowercase letters to small caps.

  - **Superscript**   Choose this style to superscript the text. When you apply this style, the characters are scaled to a smaller size and raised above the baseline.

  - **Subscript**   Choose this style to subscript the text. When you apply this style, the characters are scaled to a smaller size and shifted below the baseline.

  - **Underline**   Choose this style to underline text.

  - **Strikethrough**   Choose this style to format the text as strikethrough characters.

- **Alignment buttons**   You use the Alignment buttons to align multiple lines of text. You can have the following options for horizontal text: Align Left, Align Center, Align Right, or Justify Last Left; for vertical text: Align Top, Align Center, Align Bottom, or Justify Last Top.

- **Anti-alias option**   When you choose an anti-alias option, Adobe Encore DVD blends the edges of the text with the surrounding pixels to avoid jagged edges. You can choose from the following anti-alias options:

- ■ **None**   Anti-aliasing in not applied to text.

- ■ **Sharp**   Minimal anti-aliasing is applied to text. This setting is excellent for large text.

- ■ **Crisp**   The default setting applies stronger anti-aliasing to text.

- ■ **Strong**   Applies more anti-aliasing, and the characters appear more to have more substance. This setting works well on small text.

- ■ **Smooth**   Creates a smooth transition between text and surrounding pixels of color.

When you use the Vertical Text tool to create a text object, the alignment options in the Character palette are different as noted in the previous section. Instead of aligning text from left to right, you now align text from top to bottom, and the Character palette alignment buttons appear as shown next:

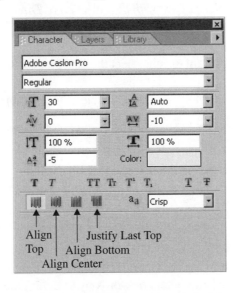

8

TIP *If you don't own Adobe Photoshop, you're probably not familiar with Pro fonts. These stylish fonts can be used to create artistic text for your menu titles, and button text. You'll be happy to know that Pro fonts are available with Adobe Encore DVD. Next time you create a new menu, peruse the Font drop-down menu in the Character palette. You'll find some interesting fonts that are not available with any other application on your system—unless you have other Adobe graphics applications installed. Try the Adobe Caslon Pro, Chaparral Pro, Adobe Garamond Pro, or Minion Pro next time you need to create an eye-catching menu title.*

Select and Edit Text

After you create text for your DVD menus, you can edit a single character, word, or an entire block of text. You can also move a block of text to a different position. You have the following options for selecting text that you want to move:

■ To select text that is not embedded in a layer set, select the Selection tool, and click the text object.

■ To select text that is part of a layer set, select the Direct Select tool, and then click the text object.

After selecting text using either of the previous methods, you can delete it by pressing DELETE or by choosing Edit | Cut. You can move the text by clicking and dragging it to another location.

TIP   *Double-click a block of text with the Selection or Direct Select tool to activate the Text tool.*

To select text that you want to edit, select the Text tool and do one of the following:

■ Click inside a text block and drag to select one or more characters.

■ Double-click a word to select it. Shift-click contiguous words to add them to a selection.

■ Place your cursor before the first character in a text block, and then shift-click the last character in a text block to select all characters in-between.

■ Place your cursor inside the text block and the press CTRL-A to select all characters.

After you select text, open the Character palette to change text characteristics. You can delete individual characters or words by selecting them using one of the methods outlined previously and then pressing DELETE or Edit | Cut. Alternatively, you can press the BACKSPACE key to delete selected text.

Add a Drop Shadow to Objects

Even though the drop shadow seems to pop up almost everywhere, it's an effective way to make text and other objects stand out from a background. If you have some particularly dainty text associated with a button, a drop shadow makes the text more prominent and easier to read. You can add a drop shadow to text or menu object while editing the menu in the Menu Editor. To add a drop shadow to text or a menu object:

1. Select the object to which you want to apply the drop shadow.

2. Choose Object | Drop Shadow. The Drop Shadow dialog box shown next appears:

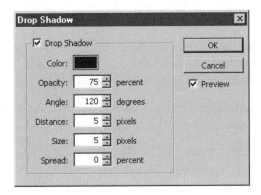

3. Click the swatch in the Color field and then choose a color from the Color Picker, shown in the following illustration. You can select a color by dragging inside the Color Picker or by entering values in the H, S, B fields to match a color from the HSB color model, or R, G, B fields to match a color from the RGB color model.

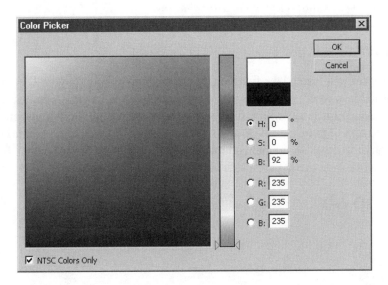

NOTE   *You can constrain the Color Picker to display only NTSC colors by clicking the NTSC Colors Only checkbox at the lower-left corner of the Color Picker.*

4. In the Opacity field, enter a value between 0 and 100. Specify a low value for a faint drop shadow or a high value for a prominent drop shadow. Alternatively, you can click the spinner buttons to specify a value.

5. In the Angle field, enter a value between 0 and 360. This determines the angle from which the shadow light source shines. Alternatively, you can click the spinner buttons to specify a value.

6. In the Distance field, enter a value. This value is measured in pixels and determines how far from the object the shadow begins. Alternatively, you can click the spinner buttons to specify a value.

7. In the Size field, enter a value. This is the size of the drop shadow in pixels. Alternatively, you can click the spinner buttons to specify a value.

8. In the Spread field, enter a value. This value is the percentage the shadow spreads beyond the object. Alternatively, you can click the spinner buttons to specify the value.

9. Click OK to apply the drop shadow to the selected object.

TIP   *To delete a drop shadow, select the object to which the drop shadow is applied and then choose Object | Drop Shadow. After the Drop Shadow dialog box opens, click the Drop Shadow checkbox. Click OK to close the Drop Shadow dialog box and remove the drop shadow.*

Summary

In this chapter, you learned the first step for creating navigation for your DVD projects. You learned how to create menus that are commonly used as main menus for DVD projects. You learned to work with Library preset menus and how to use the Text tool to add text to your menus. You also learned how to format and edit text using the Character palette. In the next chapter, you'll learn to create submenus for your projects, work with menu buttons, create links for your menu items, and more.

Chapter 9

Work with Submenus and Buttons

How To...

■ Add submenus to projects

■ Work with buttons

■ Create button links

■ Modify timeline end actions

■ Modify objects

After you create the main menu for a project, it's time to add any additional menus that link to the main menu, which are known as *submenus*. In this chapter, you'll learn to work with preset menus from the Adobe Encore DVD Library. You'll learn to specify ending actions for timelines and menus, which timeline links to a button, as well as specify links for the DVD remote control Title and Menu buttons.

Whether you use menus and submenus from the Adobe Encore DVD Library or create your own menus and submenus in Adobe Photoshop, you'll have to manipulate the menu objects in order to align and distribute them to the title safe area. You'll learn to use the Selection and Direct Select tools to accomplish these tasks, and also how to scale and move items.

Add Submenus to a Project

After you create a main menu for your project, you add submenus. Each main menu button is linked to a submenu. Furthermore, each submenu contains enough buttons for the number of timelines and/or chapter points that will be linked to the submenu. If you've planned your project ahead of time, you know how many submenus you will need and how many buttons are needed on each submenu.

If you create a submenu in Adobe Photoshop for your DVD project, create enough buttons to accommodate the submenu in your project that will contain the largest number of timelines and/or chapter points. As an alternative, you can choose a preset menu from the Adobe Encore DVD Library with a background that matches the main menu in your project. If the submenu doesn't contain enough buttons, you can always duplicate the existing buttons and then size and align them to suit your project. Alternatively, you can duplicate the main menu and modify the main menu buttons, or delete the buttons and add different buttons from the Adobe Encore DVD Library. To add a submenu to your project, follow these steps:

1. Choose Window | Menus to open the Menus tab. Alternatively, you can click the Menus tab if it's currently displayed in the workspace.

2. Choose Window | Library.

3. In the Library, click the Show Buttons and Show Images buttons to temporarily hide these items.

4. Drag the desired submenu into the Menus tab, or click the New Menu button at the bottom of the Library palette. Alternatively, you can choose File | Import as Menu to import a PSD file you've created in Adobe Photoshop. After doing either, Adobe Encore DVD opens the submenu in the Menu Editor, as shown next.

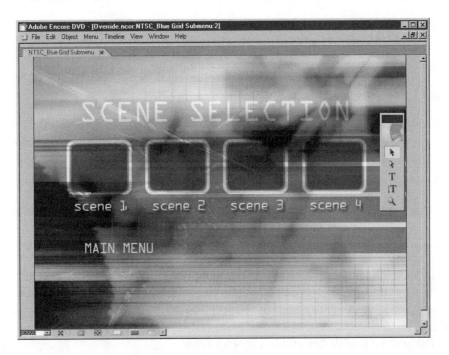

After adding a submenu to your project, you should give it a unique name to reflect its purpose in your project. You can name a submenu by right-clicking it in the Menus tab and then choosing Rename from the shortcut menu. Alternatively, you can rename the menu using the Properties palette as discussed previously in Chapter 8.

Sometimes you'll need additional buttons for your submenu; for example, a button that you'll use to navigate to the next and previous menus. You can also add a button that functions as a navigation device to return viewers to the main menu.

Add Buttons to Menus and Submenus

If you create a custom menu in Adobe Photoshop, chances are you've included enough buttons to link to the maximum number of timelines you'll use on any given project menu and to perform additional tasks such as navigating to the next or previous menu in your project. If, however, you use a menu from the Adobe Encore DVD Library, you may need to customize the menu to suit your project. For example, the Library has a button with a blue square graphic with a right-pointing arrow inside and adjacent text, which can be used for a next button. You can add a button at any time by following these steps:

1. Open the Menus tab as outlined previously.

2. Double-click the menu to which you want to add the button to open the menu in the Menu Editor.

3. Choose Window | Library to open the Adobe Encore DVD Library.

4. Click the Image and Menu buttons to temporarily hide these items.

5. Drag the desired button from the Adobe Encore DVD Library to the menu. Alternatively, you can select the desired button and then click the Place button near the bottom of the Library palette.

6. From within the Menu Editor, use the Selection or Direct Select tool to position and align the button as desired.

Working with Menu Objects

After you begin fleshing your project with navigation menus, you inevitably end up having to arrange the objects that comprise your menus. For example, when you add a button to a menu, you may need to resize the button, create duplicates of the button, and then align all menu buttons relative to the title safe area of the menu. You can also add images that you've imported as project assets to any menu.

Add Project Assets to Menus

If you've added still images as project assets, you can use them as backgrounds for your DVD menus or as a button object. To add a still image to a menu, follow these steps:

1. Choose Window | Menus to open the Menus tab.

2. Double-click the menu that you want to edit. The menu appears in the Menu Editor window.

3. Click the Project tab.

4. Select the desired image asset from the Project tab and drag it to the Menu Editor.

After adding an asset to a project, you can then position and size the object as needed using the Selection or Direct Select tool.

Use the Selection Tool

You use the Selection tool to select an object or button set. A button is an example of a layer set; the button graphic, text, and subpicture are on different layers, yet the button set is a single object. To select an object with the Selection tool:

1. Open the Menus or Project tab.

2. Double-click the menu that contains the object you want to modify.

3. Click the Selection tool to activate it, as shown here:

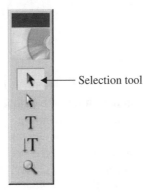

Selection tool

4. Click the object you want to select. Adobe Encore DVD displays a bounding box with eight handles around the selected object. You can now move, resize, or align the selected object as outlined in upcoming sections of this chapter.

NOTE *To select multiple items with the Selection tool, hold down the CTRL key while clicking the additional items you want to add to the selection.*

Use the Direct Select Tool

If you need to edit items that are nested within button sets or layer sets, you can easily do so using the Direct Select tool. To select an object nested within a button or layer set, follow these steps:

1. Open the Menus or Project tab.

2. Double-click the menu that contains the object you want to modify.

3. Click the Direct Select tool (shown next) to activate it:

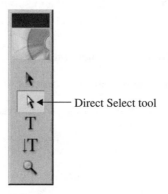

Direct Select tool

4. Click the object you want to select. Adobe Encore DVD places a bounding box with eight handles around the object you have selected. A solid bounding box appears around the entire button or layer set as well.

 You can also use the Direct Select tool to select items that are not nested in a layer set or button set.

Additionally, you can use the Direct Select tool to marquee select multiple objects by clicking and then dragging diagonally. As you drag a bounding box with dashed lines appears to signify the selection area. A bounding box with eight handles appears around each selected object. Release the mouse button when a bounding box appears around each.

Use the Zoom Tool

When you are working on a menu, you may find it necessary to zoom in on certain objects in the menu. You can easily do so with the Zoom tool. To use the Zoom tool, follow these steps:

1. Open the Menus or Project tab.

2. Double-click menu that contains the object you want to modify.

3. Click the Zoom tool, shown next, to activate it:

Zoom tool

4. Click inside the Menu Editor to zoom to the next highest level. Alternatively, you can click and drag around the area that you want to magnify.

TIP

Press the ALT key while you have the Zoom tool selected, and then click inside the menu to zoom out to the next lowest level of magnification. Release the ALT key and when you click inside the menu the tool will once again zoom in.

Modify Menu Objects

As you edit a menu, you may find it necessary to modify the objects in your menu. You can move, resize, align, and distribute items within a menu. To modify an object, you must first select it. You can select objects by clicking them with the Select or Direct Select tool, or you can select an object by clicking its name in the Layers palette.

Scale Objects

You can scale any menu object. You can scale individual objects or several objects, such as all the buttons within a menu. You can change the width, height, or both dimensions by doing the following:

1. Open the Menus or Project tab and double-click the menu you want to edit. The menu appears in the Menu Editor.

2. Select the object(s) you want to modify with the Selection or Direct Select tool. Adobe Encore DVD creates a bounding box with eight handles around each selected object. To scale the object(s), do one of the following:

    ■ Click the top or bottom handle and then drag up or down to change the height of the object(s).

    ■ Click the middle right or middle left handle and then drag left or right to change the width of the object(s).

    ■ Click one of the corner handles and then drag diagonally to change the width and height of the object disproportionately.

    ■ Click one of the corner handles and then drag diagonally while holding down the SHIFT key to resize the object proportionately.

3. Release the mouse button when the object is the desired size.

NOTE   *When resizing multiple objects, dragging a handle on any bounding box applies the transformation to all selected objects.*

Duplicate Objects

After you add a button or other object to your menu, you can duplicate it as needed. This is especially useful if you begin with a blank menu and then flesh it out with preset buttons from the Adobe Encore DVD Library. You can duplicate objects using menu commands or by using the Selection or Direct Select tool. To duplicate an object, open the Menus or Project tab and then double-click the menu you want to edit. After the menu appears in the Menu Editor do one of the following:

■ Select the desired object and then choose Edit | Duplicate. After choosing this command, Adobe Encore DVD creates a carbon copy of the selected object to the right and slightly below the object.

■ Select the desired object and choose Edit | Copy. Deselect the object and then choose Edit | Paste. Adobe Encore DVD pastes the copied object into the menu. Alternatively, you can press CTRL-C to copy the object, and CTRL-V to paste it.

■ Select the desired object, press the ALT key, and then drag. Release the mouse button to create a duplicate. After you begin dragging, press the SHIFT key to constrain motion to vertical, horizontal, or 45-degree diagonal.

TIP   *You can copy an object from one menu and paste it into another.*

Delete Objects

As you build a DVD menu, you may find a button is not needed; for example, when you duplicate a submenu and don't need as many buttons as the original. You can select one or more items and then delete them by doing one of the following:

■ Choose Edit | Cut. This command cuts the selected items, but they remain on the system clipboard.

■ Choose Edit | Clear. This command deletes the selected items from the menu.

■ Press the DELETE or BACKSPACE key.

TIP   *If you delete an item in error, choose Edit | Undo or press CTRL-Z.*

Arrange Objects

When you add objects to a menu, they are stacked in the order in which you add them. The last item added appears at the top of the stack and eclipses items that are underneath it and lower in the stacking order. You can change the appearance of a menu by rearranging the stacking order of objects. To change the stacking order of an object:

1. Select the object whose stacking order you want to change.

2. Choose Object | Arrange, and then choose one of the following from the submenu:

9

■ **Bring to Front**   Moves the selected object to the top of the stacking order.

■ **Bring Forward**   Moves the selected object above the next object in the stacking order.

■ **Send Backward**   Moves the selected object underneath the next object in the stacking order.

■ **Send the Back**   Moves the selected object to the bottom of the stacking order.

Align Objects

After adding buttons and other objects to a menu, you need to align them. You can align two or more objects relative to the position of the selected objects, or you can align one or more objects relative to the title safe area.

To align objects relative to each other, follow these steps:

1. Select the objects you want to align.

2. Choose Edit | Align and then choose one of the following options:

   ■ **Left**   Aligns selected objects to the leftmost object in the selection.

   ■ **Center**   Aligns selected objects to the relative center between the two objects.

   ■ **Right**   Aligns selected objects to the rightmost object in the selection.

   ■ **Top**   Aligns selected objects to the topmost object in the selection.

   ■ **Bottom**   Aligns selected objects to the bottommost object in the selection.

To align one or more objects relative to the title safe area, follow these steps:

1. Select the object(s) you want to align.

2. Choose Object | Align | Relative to Safe Area. This command notifies Adobe Encore DVD that the upcoming align commands will be relative to the title safe area of the menu.

3. Choose Object | Align and then choose one of the following:

- ■ **Left**   Aligns selected objects to the left border of the title safe area.

- ■ **Center**   Aligns selected objects to the center of the title safe area.

- ■ **Right**   Aligns selected objects to the right border of the title safe area.

- ■ **Top**   Aligns selected objects to the top border of the title safe area.

- ■ **Middle**   Aligns selected objects to the middle of the title safe area.

- ■ **Bottom**   Aligns selected objects to the bottom of the title safe area.

NOTE   *The Relative to Safe Area option remains in effect until you choose Object | Align | Relative to Safe Area again.*

Distribute Objects

If you need to evenly space buttons or other objects, you can do so easily with menu commands. You can distribute items relative to the position they occupy within the menu when you select three or more objects or distribute two or more objects relative to the title safe area. Objects are distributed relative to their center point.

To distribute objects relative to their current position, follow these steps:

1. Select three or more objects within a menu. Note that if the objects are text or button sets, they should reside within the title safe area; if the objects are graphic images, they should reside within the action safe area.

2. Choose Object | Distribute and then choose one of the following from the submenu:

- ■ **Vertically**   Distributes objects equally along the vertical axis of the menu between the topmost object and the bottommost object.

- ■ **Horizontally**   Distributes objects equally along the horizontal axis of the menu between the leftmost object and the rightmost object.

9

To distribute objects equally relative to the title safe area of the menu, follow these steps:

1. Select two or more objects within a menu.

2. Choose Object | Distribute | Relative to Safe Area.

3. Choose Object | Distribute, and then choose one of the following from the submenu:

   ■ **Vertically**   Distributes selected objects equally between the top and bottom of the title safe area.

   ■ **Horizontally**   Distributes selected objects equally between the left and right border of the title safe area.

Rotate and Mirror Objects

Adobe Encore DVD 1.0 has no tool or menu command to rotate or mirror an object. If you need to rotate or mirror an object (for example, an arrow image from the Adobe Encore DVD Library), you can do so by choosing Menu | Edit in Adobe Photoshop. After the menu opens in Adobe Photoshop, you can use the Free Transform command to rotate an object, or you can rotate or mirror the object by using one of the commands

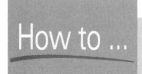 **Rotate and Mirror an Object Without Adobe Photoshop**

If you don't own Adobe Photoshop, you may be able to rotate or mirror the object in another image editing application by selecting the object in Adobe Encore DVD and then choosing Edit | Copy or by pressing CTRL-C. Launch your image editing application, create a new document, and then choose the application's Paste command or press CTRL-V. Rotate or mirror the object using your image editing application's tools and menu commands, and then press CTRL-C to copy the edited object to the system clipboard. Within Adobe Encore DVD, choose Edit | Paste to paste the edited object into your menu. This may not work will all image-editing applications. Before trying this technique on an important project, create a new project, add a graphic blank menu, and then try to copy and paste it into another image-editing application.

from the Transform submenu. After rotating the object, save the menu in Adobe Photoshop, and your edits will be applied to the menu in Adobe Encore DVD.

Convert Objects to Buttons

If you import images as project assets, you can also use them as buttons. You can also convert text objects to buttons. For example, if a client has a unique company logo for each division of their company, they may request that you use them as graphics for buttons that link to video clips or slide shows about each division. To convert a graphic object to a button, follow these steps:

1. Click the Project tab and then double-click the menu you want to edit. The menu opens in the Menu Editor.

2. Select the object that you want to convert to a button.

3. Choose Window | Layers to open the Layers palette. The object's name is highlighted in the palette.

4. Click the object's Button/Object column. Adobe Encore DVD adds a button icon to the object's Button/Object column.

9

TIP   *You can also convert an object to a button by selecting it in the Menu Editor, then choosing Object | Convert To Button.*

Add Text Subpictures

When you convert a text object to a button, by default, it does not display a subpicture when selected. You can create a subpicture for text menu objects from within Adobe Encore DVD by following these steps:

1. Select the text object that you've converted to a button.

2. Choose Object | Create Subpicture. Adobe Encore DVD adds a highlight layer to the applicable button set using the menu's color set. Alternatively, you can open the Properties palette and click the Create Text Subpicture checkbox.

Create a Menu Template

Whether you create a menu in Adobe Photoshop or assemble one in Adobe Encore DVD using the Library menu presets and imported assets, chances are you will use a variation of the menu throughout a DVD project. In this regard, you can use the first iteration of the menu as a template and duplicate it for the other menus and submenus your project requires. If you've meticulously planned your project, you'll know exactly how many menus are required.

Duplicate Menus

After you create your first menu for a project, you can duplicate the menu as needed to create enough menus to display each timeline and chapter point in your production. You can duplicate a menu from within the Menus or Project tab. To duplicate a menu, follow these steps:

1. Open the Menus or Project tab.

2. Select the menu you want to duplicate.

3. Choose Edit | Duplicate. Adobe Encore DVD creates a duplicate of the menu and assigns it the name of the parent menu appended by the word *copy*.

4. Repeat Step 3 to create enough menus for your project. As additional menus are created, they are appended with *Copy 1, Copy 2,* and so on.

After you create enough menus for your project, you can edit each menu to suit it for the chapter points to which it will link by editing title text and adding or deleting buttons as needed. You should also rename your duplicate menus at this point, as this will simplify the job of creating links from the main menu to submenus and from timelines back to menus.

 You can delete an unwanted menu at any time by selecting it in the Menus or Project tab and then choosing Edit | Clear. Alternatively, you can press DELETE or BACKSPACE.

Create Menu and Button Links

After creating the needed number of menus, you create links from menus to timelines and vice-versa. In addition to creating menu links, you also need to create a link from the timeline back to a menu. This is known as an *end action.* You can create links using the Properties palette, by dragging and dropping timelines or timeline chapter points into existing menus, or by using a clever device known as the *pick whip*. In the upcoming sections, you'll learn how to create links. Each section will show a different method for creating a link. After you gain experience with Adobe Encore DVD, you'll end up with a favorite method for creating links; however, each method is useful for creating certain links.

Create Menu Links

Many DVD authors create a main menu that lists the options from which viewers can choose. A main menu is typically comprised of text objects that have been converted to buttons. If the DVD is a movie, the main menu can have a button to play the entire timeline and other buttons that link to submenus. Each submenu has buttons that link to individual scenes, offer audio options, subtitle options, and so on. To set menu links follow these steps:

1. Click the Project or Menus tab.

2. Double-click the menu whose links you want to set to display the menu in the Menu Editor.

3. Select the button you want to set.

4. Choose Window | Properties to open the Properties palette.

5. Click the triangle to the right of the Link field and choose a link from the drop-down menu, as shown next. This menu lists up to 20 of the most recently used menu and timelines.

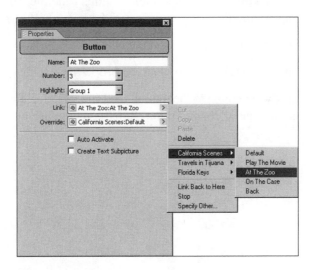

6. Click Specify Other to open the Specify Link dialog box if the desired menu or timeline is not shown.

7. Choose the desired menu or chapter point and click OK to exit the dialog box and set the link for the button.

When you add a link to a button, it appears in the Properties palette, as well as on the button's Link column in the Menus tab. You can also manually set a menu button link using the pick whip, a technique that will be covered in the next section.

Link Menu Buttons to Chapter Points

If you've already created a menu with buttons, you're ready to link them to a timeline or chapter points within a timeline. Again, there are a number of ways you can accomplish this task. In this section, you'll learn one of the easier ways to create a link using the pick whip. To link a menu button to a timeline or chapter point, follow these steps:

1. Open the Menus or Project tab and double-click the menu for which you want to create button links. This opens the menu in the Menu editor.

2. Minimize and resize the Menu editor and Project window so that you can freely move from one to the other. If you have enough desktop space, align the two windows so they are side by side. Make sure that you have access to all menu buttons.

3. Click the Timelines tab. All project timelines are displayed at the top of the tab. If you're linking to a chapter point within a timeline, click the timeline name to display timeline chapter points in the lower half of the tab.

4. Select the desired button in the Menu editor.

5. Choose Window | Properties to open the Properties palette.

6. Click the pick whip button to the right of the Link field and then drag towards the desired timeline or chapter point. Release the mouse button and the link is set.

NOTE   *If you link to a timeline with multiple chapter points, the link is to the first chapter point in the timeline.*

Link Timelines to Menus

If you work with blank menus or a menu to which you have not added buttons, you can flesh out a menu complete with linked buttons in less than a minute. You do so by working with the Timelines tab and Menu editor open. To link timelines to a menu while simultaneously creating buttons, follow these steps:

1. Open the Menus or Project tab.

2. Double click the menu for which you want to create links. Adobe Encore DVD displays the menu in the Menu editor.

3. Click the Timelines tab to display all of your project timelines.

4. Arrange and size the Timelines tab and Menu editor so they're both visible in the workspace. If you have enough desktop space, arrange the windows so they're side by side. Note that you may have to reduce the magnification of the menu in order to view its entirety.

5. Select the timeline you want to link to the menu and drag it into the Menu editor. As your cursor crosses the Menu editor, a square with a curved arrow appears beneath the cursor icon.

6. Release the mouse button. Adobe Encore DVD links the timeline to the menu and creates an instance of the default Library button. A video thumbnail of the timeline poster frame appears inside the button.

7. Repeat Steps 5 and 6 as needed to add additional links to the menu.

9

TIP
You can also use this technique to edit menu button links. Drag a timeline to an existing menu button. When you release the mouse button, the previous link is overridden. You can also drag a timeline to a button for which a link has not been previously specified.

Figure 9-1 shows a menu that has been fleshed out by dragging timelines into a blank menu. To create the finished menu shown in Figure 9-1, the button text was edited and the buttons were aligned as discussed previously in this chapter.

Create Menu Links to Timelines with Multiple Chapter Points

The previous technique works wonderfully when you have only one chapter point per timeline. When you have multiple chapter points on a timeline, you need to create

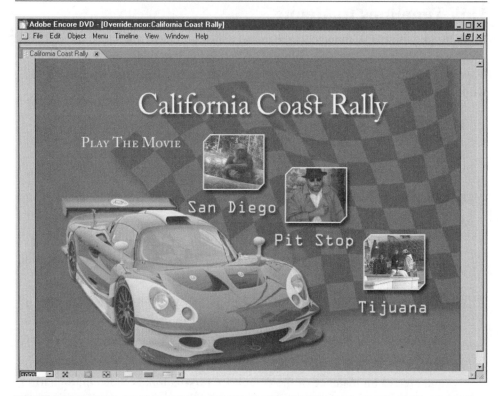

FIGURE 9-1   You can flesh out a menu by dragging timelines into a menu being edited in the Menu editor.

a link for each chapter point. You can do so by adding buttons to a menu and then specifying the link in the Properties palette, as outlined previously. However, this is time consuming. You can quickly create links for a timeline with multiple chapter points by following these steps:

1. Open the Menus or Project tab.

2. Double click the menu for which you want to create links. Adobe Encore DVD displays the menu in the Menu editor.

3. Click the Timelines tab to display all of your project timelines.

4. Arrange and size the tabs so they're both visible in the workspace.

5. Select the timeline whose chapter points you want to link to a menu.

6. Select a chapter point from the bottom window and drag it into a blank area of the menu to create a new button and link simultaneously, or drag the chapter point to an existing menu button to link the button to the chapter point.

7. Continue dragging timeline chapter points into the menu to finish setting links.

Set Default Library Button

When you drag a timeline or chapter point into a blank menu or a blank area of a menu, Adobe Encore DVD inserts the Library's default button, simultaneously linking the button to the timeline or chapter point. If desired, you can set a different button as the default Library button by following these steps:

1. Choose Window | Library. The Adobe Encore DVD Library appears.

2. If desired, temporarily hide Library images and menus by clicking the Show/Hide Images and Show/Hide Menus buttons.

3. Select the button you want to set as Library default.

4. Right-click and choose Set as Default Button from the shortcut menu. After choosing this command, two asterisks (**) appear before the button, signifying it is the default button. Adobe Encore DVD will insert this button into a menu anytime you drag a timeline or chapter point into a blank menu or blank area of a menu.

Set Default Menu Button

When you create a menu and add buttons to a menu, Adobe Encore DVD sets the first button you create as the default button. Default appears as one of the options for a link, end action, or override in the Properties palette drop-down list for the menu. A menu's default button is highlighted when a viewer selects the menu. You can set the default menu button by following these steps:

1. Open the Menus or Project tab.

2. Select the desired menu from the list.

3. Choose Window | Properties.

4. Click the triangle to the right of the Default Button field and select the desired button from the drop-down list, shown next:

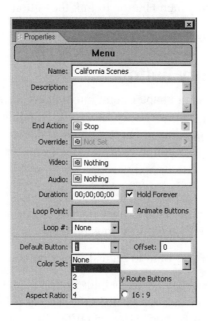

Renumber Buttons

When you add buttons to a menu, Adobe Encore DVD automatically numbers the buttons. Button numbers appear on the Number drop-down list when a button's properties are displayed in the Properties palette. Button numbers are displayed when you show button routing for a menu and are used when setting the default menu button. When you change the position of menu buttons, Adobe Encore DVD does not renumber the buttons. You can, however, manually renumber buttons by following these steps:

1. Open the Menus tab.

2. Double-click the menu whose button numbering you want to change. The selected menu opens in the Menu Editor.

3. Click the Show Button Routing button at the bottom of the editor. Each button's number is displayed in the center of the routing arrows.

4. Select the desired button.

5. Choose Window | Properties to open the Properties palette. The button's current number is displayed in the Number field.

6. Click the triangle to the right of the Number field and select the desired number from the drop down list. After you specify a button number, the other menu buttons are renumbered and the button routing numbers are updated in the Menu Editor.

7. Repeat Steps 3–6 to renumber additional buttons.

You can also renumber buttons by selecting a menu in the Menus tab, selecting the button from the lower half of the Menus tab, and then using the Properties palette to select the desired button number. After selecting a different button number, each menu buttons' Number column is updated to reflect the new button numbering sequence for the menu.

Set Button to Auto Activate

As a rule, when you author a DVD, you set a button so that the link is played when a viewer presses the ENTER button on his or her DVD remote controller. You can, however, set a button to auto-activate when the viewer selects the button. This feature can be used to create hidden "Easter Eggs" (hidden surprises from program or menu designers) in the form of different versions of the same menu that display upon when the button is selected, or if a user passes a mouse over the button when the DVD is viewed on a computer. To set a button to auto activate, follow these steps:

1. Open the Menus or Project tab.

2. Double-click the menu that contains the button whose link you want to set. Adobe Encore DVD displays the menu in the Menu Editor.

3. Select the button you want to auto-activate. Note that you can also set a button to auto-activate by clicking the menu in the Menus tab and then selecting the button in the lower-half of the Menus tab.

4. Choose Window | Properties to display the Properties palette.

5. Set the button link as outlined previously. The link can be to another menu or a timeline.

6. Click the Auto-Activate checkbox.

9

Set Video Timeline End Action

The default action for a video timeline is Stop. If you accept this default end action, the television screen will go black when the timeline finishes playing. You can change the end action to a different outcome, such as playing another video timeline or displaying the menu from which the timeline was called. To set a link for the end action of a video timeline, follow these steps:

1. Open the Project or Timelines tab.

2. Select the timeline for which you want to set the end action.

3. Choose Window | Properties to display the Properties palette.

4. Click the triangle to the right of the End Action field and choose a link from the drop-down list. Alternatively, if you are currently editing a menu in the Menu Editor, you can drag the End Action pick whip to the desired menu button.

Set Override Action

When you set an end action, you control the DVD viewer's experience by specifying a link to menu or timeline. For example, if you were creating a DVD for a movie, you would create a Play Movie button that links to the first scene of the movie. When you create a DVD of this type, you would create a link to the next scene's timeline for the end action of each scene.

However, you also want to give your viewers some flexibility in case they want to view an individual scene. To do this, you would create a submenu for a specified number of scenes and then create a button that links to each scene's timeline. You would then link each button to the applicable timeline and set the button's override action to link to the menu. When you set an override link for a button, this link overrides the timeline's end action and returns the viewer to the menu. To set an override action while editing a menu, follow these steps:

1. Select the button whose Override action you want to set.

2. Choose Window | Properties to display the Properties palette.

3. Click the Override Action field pick whip and drag it to the default menu button, or if desired, the next button in the menu.

4. When the button is highlighted, release the mouse button to set the link.

Set Remote Title Button Link

Most set top DVD player remote controllers are equipped with a button that returns the DVD presentation to the main menu. DVD remote controllers label this button as Top Menu or Title. You can designate which menu plays when the Top Menu button is clicked by following these steps:

1. Choose Window | Disc to display the Disc tab.

2. Choose Window | Properties to display the Properties palette, shown next:

3. Click the triangle to the right of the Title Button field and choose the menu to which the title button links from the drop-down menu.

4. Choose Specify Other to display the Specify Link dialog box if the desired menu is not listed.

5. Choose the menu to which the Title Button links and then click OK.

When you set a title button link, it is not usually necessary to specify to which title menu button the link is—in most cases you want to link to the default button of the title menu.

9

Set Remote Menu Button Link

Most set top DVD player remote controls have a Menu button. When viewers are watching a selection, they can click the Menu button to cease playing the selection, return to the menu, and select another track. You can set the Menu Remote link by following these steps:

1. Open the Timelines tab.

2. Select the desired timeline.

3. Choose Window | Properties to open the Properties palette.

4. Click the triangle to the right of the Remote Menu link and choose the desired menu from the drop-down list. Alternatively, if the applicable menu is open in the Menu editor, you can drag the pick whip to the desired menu button and release the mouse button to set the link. In most cases, you should link to the default menu button.

Preview a Menu

You can test menu links for the entire project before building a disk. You can also preview a menu to test button links and linked timelines to make sure the end actions execute as desired. If you're working on a complex project, it may be advantageous to check a menu's button links before moving on to the next phase of your project. To preview a menu you're currently editing in the Menu editor, follow these steps:

1. Right-click anywhere in the Menu editor and choose Preview from Here from the shortcut menu. Adobe Encore DVD displays the menu in the Project Preview window, as shown in Figure 9-2.

2. Use the Remote Control buttons at the bottom of the window to select and activate buttons. When you activate a button, the linked timeline will play. These buttons function identically to the remote control buttons on a set top DVD player.

As you preview the menu, make sure your timeline end actions execute as desired. If you don't have time to watch an entire timeline or chapter, you can click the End Action button to execute the end action before the timeline or chapter finishes playing. You can select and activate menu buttons, as well as select alternate sound tracks, subtitle tracks, and more. Figure 9-2 shows the buttons you use to test a menu and its buttons. You'll find detailed discussion of every feature of the Project Preview window in Chapter 14.

Title Remote button
Menu Remote button | End Action button
DVD Remote Control buttons

FIGURE 9-2   You can test button links by previewing a menu.

Edit Menu Links

If after previewing the menu you find that some of your links or timeline end actions are in error, you can edit them by selecting the button or timeline and then opening the Properties palette. From within the Properties palette, you can specify a different link, end action, and so on, as outlined in the previous sections of this chapter.

Set Remote Control Button Routing

When you add buttons to a DVD menu, Adobe Encore DVD automatically routes the buttons. Button routing determines the order in which buttons are selected when viewers select the arrow buttons on their DVD remote controllers. You can specify the order in which Adobe Encore DVD routes buttons in the Menus section of the Preferences dialog box as discussed previously in Chapter 3. If desired, you can also manually set the routing order of menu buttons.

View Default Button Routing

Before you can set the routing order of buttons, you invoke a menu command to display button routing while editing a menu. To display button routing, follow these steps:

1. Open the Menus or Project tab.

2. Double-click the desired menu to open it in the Menu Editor.

3. Choose View | Show Button Routing. Alternatively, you can click the Show Button Routing icon at the bottom of the Menu Editor. Adobe Encore DVD displays a graphical representation of button routing, as shown in Figure 9-3.

In the center of the button routing display is each button's number. The arrows surrounding the button number indicate which button is selected when the corresponding DVD remote control arrow button is clicked. If the arrow icon doesn't show the desired button routing, you can manually adjust button routing, as discussed in the next section.

Manually Set Button Routing

If the default button routing suits your project, you can go on to the next phase of menu editing. If desired, you can manually set button routing by following these steps:

1. Open the desired menu in the Menu Editor and then choose Menu | Show Button routing as outlined in the previous section. Adobe Encore DVD displays the button routing for the menu shown in Figure 9-3.

2. Choose Window | Properties to open the Properties palette.

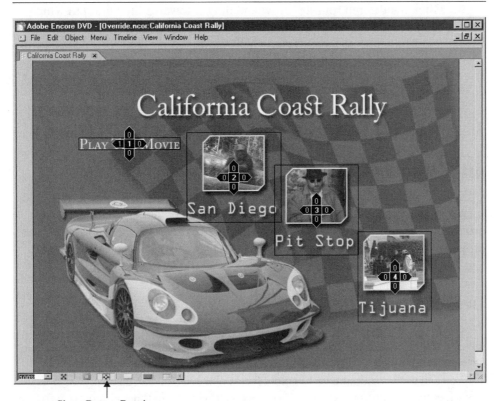

Show Button Routing

FIGURE 9-3   You can display the routing of buttons in the Menu Editor.

> **NOTE**   *Depending on the complexity or your project, and the number of menus within, you may see the properties for the menu's default button when opening the Properties palette. If this occurs, click the name of the menu you are editing in the Menus or Project tab.*

3. Click the Automatically Route Buttons checkbox to deselect the option.

4. In the Menu Editor, use the Selection or Direct Select tool to select a button you want to reroute. After selecting the button, the routing information is highlighted in blue.

5. Click the desired routing arrow and drag it towards the button to which it should route. As you drag towards another button, your cursor becomes a hand, and a line links to the arrow you are routing. When your cursor passes over another button's active area, the button is highlighted. To complete routing for the arrow, release the mouse button when the desired button is highlighted.

6. Repeat for the other arrows on the button, and then repeat for the other buttons on the menu.

About Menu Color Sets

Color sets define the colors used for button subpictures. You can specify one color set per menu. You can, however, use different color sets for other menus in your project. To achieve a cohesive look, it's desirable to use the same color set for each project menu. In the upcoming sections, you'll learn how to work with color sets and assign them to menus and timelines.

Define Menu Color Set

When you define a color set for a menu, you can use the default color set, modify the default color set, or create one of your own design. You can export a custom color set for future projects.

You can use a maximum of 15 colors for a color set. You can also vary the opacity of each color to suite your design. The colors are divided into three groups: one for the normal (when a button is not selected) state and two for the highlight (when a button is selected or activated) states. You can select up to six colors for each highlight group: three for the selected states and three for the activated states. You assign a highlight group to each button in your menu. This is useful for designating different button types on each menu. You can use one highlight group for buttons that link to timelines or chapter points and another highlight group for buttons that link to different menus; for example: the next, previous, or main menu.

To define a color set, follow these steps:

1. Choose Edit | Color Sets | Menu. The Menu Color Set dialog box appears, as shown in Figure 9-4.

2. Click the New Color Set button. The New Color Set dialog box appears.

3. Enter a name for the color set and click OK.

4. Click a color swatch you want to define. The Color Picker shown next appears.

Color well   Color Spectrum sliders

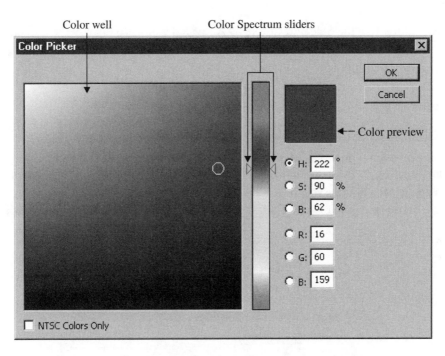

5. Click NTSC safe colors if you're authoring an NTSC DVD.

6. Specify the desired color by dragging the color spectrum sliders and then clicking inside the color well. Alternatively, you can enter values in the H, S, B fields, or the R, G, B fields. Note that if you've enabled NTSC safe colors, Adobe Encore DVD automatically replaces any non-NTSC safe color values with values that create the closest match to an NTSC safe color. The current color appears in the top of the Color Preview window while the selected color appears in the bottom half of the window.

TIP   *You can automatically select an NTSC Safe color by entering a value between 16 and 235 in the color pickers, R, G, and B fields.*

7. Click OK to set the color.

8. Click the triangle to the right of the Opacity field and choose the desired value from the drop-down list.

9. Repeat Steps 4 through 9 for other colors you want to define.

10. Click Use the Selected Colors as Activated Colors to display the Selected Colors when a button is selected or activated. This option is also available for Highlight Groups 1 and 2.

11. Click OK when you've specified colors for the set.

Understand the Default Menu Color Set

When you use the Import as Menu Asset command to import a PSD file into a project, Adobe Encore DVD creates a menu color set based on the colors you selected when you created the buttons sets in Adobe Photoshop. The subpicture colors are used to define identical colors for the selected and activated states. The normal state opacity

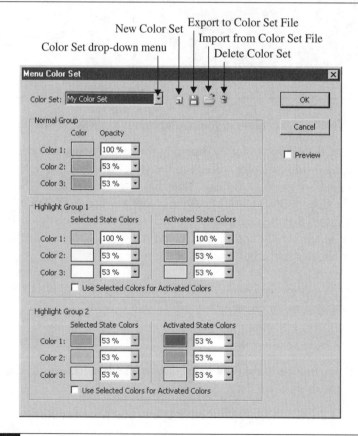

FIGURE 9-4   You can create a custom color set for your DVD projects.

is set to 0 percent. With this default color set, the highlight and select colors are the same, and menu buttons are not highlighted in the normal state.

To create the color set, Adobe Encore DVD uses the colors from the button set stacked closest to the background layer. The color set uses color information from layer (=1) of the button set for Color 1 of Highlight Group 1 and uses the color information from layer (=2) and layer (=3) for Color 2 and Color 3. If there are additional button sets in the document that contain different colors, Adobe Encore DVD uses the next two highest button sets in the stacking order to set the colors for Highlight Group 2 and Highlight Group 3.

If you edit the menu in Adobe Photoshop and change subpicture colors, Adobe Encore DVD updates the color set as soon as you save the menu file in Adobe Photoshop. You cannot edit the Automatic Color set in Adobe Encore DVD as you can with other color sets. You can, however, create a new color set based on the automatic color set. To do so, follow these steps:

1. Open the Menus or Project tab and choose the menu whose automatic color set you want to modify.

2. Choose I Edit I Color Set I Menu to open the Menu Color Set dialog box.

3. Click the New Color Set button to open the New Color Set dialog box.

4. Name the color set and click OK.

5. Specify color and opacity setting for the Normal and Highlight groups as outlined previously.

6. Click OK to save the color set.

When you create a custom color set, it is saved with the project and is not available when you create new projects. You can, however, export a color set from one project, and import it into another as outlined in the upcoming Import and Export Color Sets section.

TIP   *After creating a custom color set, you can use it with other menus in your project.*

Specify Menu Color Set and Button Highlight Group

After you create one or more color sets for a project, you can assign them to menus. After assigning a color set to a menu, you can specify which highlight group menu buttons will use. You cannot specify a highlight group when you choose Automatic as the color set option.

To select a menu color set, follow these steps:

1. Open the Menus or Project tab.

2. Select the desired menu.

3. Choose Window | Properties to open the Properties palette.

4. Click the triangle to the right of the Color Set: field and choose the desired color set from the drop-down list.

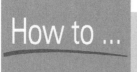

Set Properties for Multiple Objects

When you're setting a property for menu buttons or specifying actions for timelines, you often use the same link or same property for several items. For example, when you set the override action for timelines that are linked to a menu, they generally link to the menu default button. And when you specify a highlight group for menu buttons, as a rule, you use one highlight group for buttons that link to video timelines and the other highlight group for buttons that link to other project menus. You can easily set the same property for multiple items by following these steps:

1. Access the tab that contains the desired items. If you're setting properties for timelines, open the Timelines tab; if you're setting options for menu buttons, select the desired menu from the Menus tab and then select the desired buttons from the bottom half of the Menus tab, and so on.

2. Choose Window | Properties to open the Properties tab. When you select multiple items, the Name field of the Properties tab displays <<X Values>>, X being the number of items you have selected. The following illustration shows the Properties palette as it appears when three items have been selected.

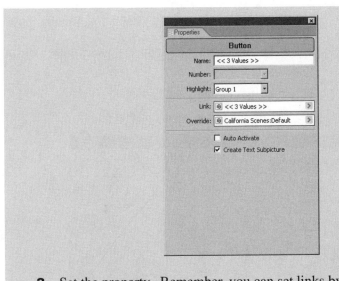

3. Set the property. Remember, you can set links by making a choice from a drop-down list or by dragging the pick whip to a timeline, chapter point, menu button, and so on.

9

After you select a color set for a menu, you can specify which highlight group each button in the menu uses. To specify a highlight group for a menu button:

1. Open the Menus or Project tab.

2. Select the desired menu.

3. Select the desired button.

4. Click the triangle to the right of the Highlight field and choose Group 1 or Group 2.

If you have multiple menus in a button, you can specify a highlight group for several buttons at once by following the steps in the above sidebar.

Import and Export Color Sets

When you create custom color sets for a DVD project, they are saved with the project and not available for other projects. When you create a color set that you may want to use with other projects, you can export the color set. After you export a color set, you can import it to use with another project's menus.

To export a color set, follow these steps:

1. Choose Edit | Color Sets | Menu to open the Menu Color Set dialog box, shown next:

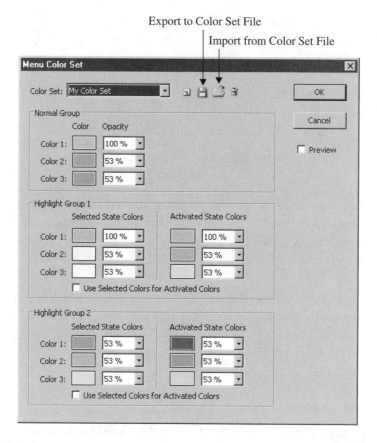

2. Click the triangle to the right of the Color Set field and choose the desired color set.

3. Click the Export to Color Set File button to open the Save Color Set File dialog box.

4. Enter a name for the color set and then navigate to the desired folder.

5. Click Save. Adobe Encore DVD exports the color set file to the specified folder. All color set files have the CS extension.

If you create a lot of color set files, it's a good idea to store them in the same folder. You should also give each color set file a unique name that will make it easier for you to identify the purpose for which you created a color set file. For example, you might name a file after the client for whom you created the DVD project or specify the type of background for which the color set is suited.

To import a color set file into a project, follow these steps:

1. Choose Edit | Color Sets | Menus to open the Menu Color Set dialog box.

2. Click the Import from Color Set file button to open the Import Color Set dialog box.

3. Navigate to the folder in which you store your color set files and then select the desired color set.

4. Click Open. Adobe Encore DVD imports the color set file, which now appears on the Color Set drop-down list in the Menu Color Set dialog box and Properties palette.

9

Summary

In this chapter, you learned to specify button links, end actions for timelines, and override actions for menu buttons. You also learned how to create custom color sets, specify which color set is used by menus, and which menu color group is used by buttons. In the next chapter, you'll learn to create motion menus for your DVD projects.

Chapter 10

Create Motion Menus

How To...

- Create video for motion menus
- Create motion menus
- Add audio tracks
- Create animated buttons
- Render motion menus

If you want to add the sizzle to your steak, so to speak, consider adding motion menus to your DVD projects. When you add motion menus, you pique viewers' interest by showing them some moving eye candy while they ponder which menu choice to select. Combine a motion menu with some background music, and you've got a professional-quality DVD production. Another way you can pique viewer interest is by animating buttons. As you know, most button sets have a video thumbnail layer that displays the chapter point poster frame. When you create an animated button, Adobe Encore DVD renders a video the size of the thumbnail. The video plays when the menu is selected.

In this chapter, you'll learn how to create motion menus. You'll also learn how to create menus that play audio tracks when activated. You'll also learn to add even more excitement to your DVD productions in the form of animated buttons. In the upcoming sections, you'll find all the necessary information to add motion menus and animated buttons to your DVD project, specify the number of times each motion menu plays, and how to set the duration of the menu's animated buttons.

Create Video for Motion Menus

If you're an accomplished videographer, you can create video for use as backgrounds for DVD menus. Armed with your digital camcorder, you can capture video of interesting cloud formations, crashing surfing, wind whipping amber fields of grain, and so on. Remember your goal is to create viewer interest. However, when you shoot your video, keep in mind that you'll have to display title and button text over the motion menu, as well as buttons. In this regard, shoot a subject that's interesting but has dark areas over which you can display text. For example, fifteen or twenty seconds of the last stages of a sunset can make for an interesting background for a motion menu. Another excellent choice would be night video city lights punctuated by flashes of lightning from a distant thunderstorm.

If you own a video-editing application that can create generated media, you can produce interesting effects such as a moving star field over a still image of planets in outer space. If you own Adobe After Effect, you can create background videos with animated text. Figure 10-1 shows a frame of a motion menu background video that was created in a video-editing application as displayed in the Adobe Encore DVD Project Preview Window.

NOTE *If you create background videos for motion menus in a video-editing application, be sure to render them in a DVD compliant format recognized by Adobe Encore DVD.*

FIGURE 10-1   You can use background videos that are created in a video editing application.

Another excellent source for background videos is the Adobe Encore DVD installation CD ROM. In the Menu Backgrounds section Goodies folder, you'll find several background videos for the NTSC and PAL formats. Be sure to read the EULA (End User License Agreement) before using these videos. Figure 10-2 shows an example of a menu background video from the installation disc.

Import a Motion Menu Video

When you create motion menus, you don't work with timelines as you do with other video assets, but you do import the video into your project as with any other asset. When you use a video clip to create a motion menu, the video clip replaces the background of the menu. All the other menu items—such as text and buttons—appear

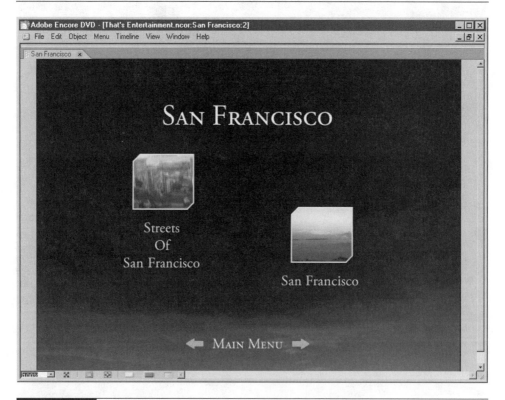

FIGURE 10-2   You can use background videos from the Adobe Encore DVD installation disc.

above the background video clip. After importing the video to your project, you use it as a background for an existing menu by following these steps:

1. Open the Project tab.

2. Select the menu you want to convert to a motion menu.

3. Choose Window I Properties to open the Properties palette.

4. Click the Video pick whip and drag it to the desired video asset.

> **TIP** *You can also set a video asset as a menu background by selecting the video asset in the Properties tab, and then while holding down the ALT key, drag the asset to the menu. When you release the mouse button, Adobe Encore DVD replaces the current background with the video.*

5. Release the mouse button. Adobe Encore DVD replaces the menu's current background with the video clip and displays the video title and the duration of the video in the Properties palette, as shown next:

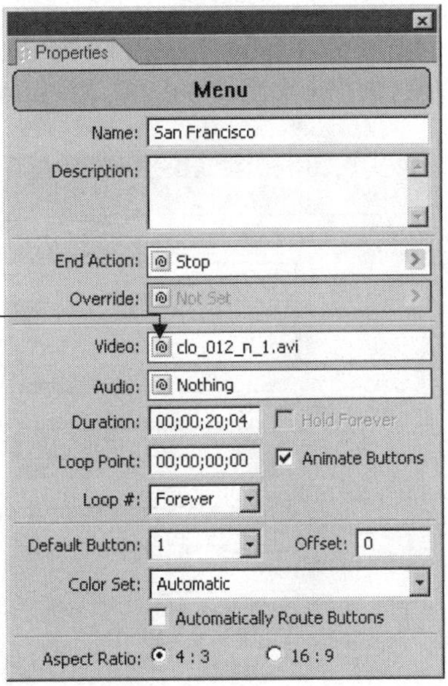

Video pick whip ——

10

Add Background Audio

When you set properties for a DVD menu, you can involve another viewer sense by adding a background audio track. You can import any sound type supported by Adobe

Encore DVD. When you add background audio to a menu, you do not have to create a timeline for the track.

Background audio plays for the duration of the menu and the number of loops you specify. If the audio is not of sufficient duration, there will be silence until the menu loops again. For more information on menu looping, refer to the previous section. To add a background audio track to a menu, follow these steps:

1. Import the audio track you want to use as background menu audio.

2. Open the Project tab.

3. Select the desired menu.

4. Choose Window | Properties to open the Properties palette.

5. Click the Audio pick whip and drag it to the desired audio track.

TIP   *You can also set a background sound for a menu by selecting it in the Project tab and then dropping it into the menu while you are editing it in the Menu Editor.*

6. Release the mouse button. Adobe Encore DVD displays the audio filename in the Audio field of the Properties palette, as shown next:

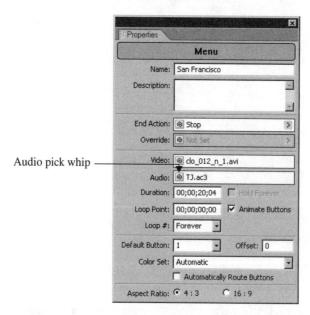

How to ...   Create Background Audio Tracks

You can create your own audio tracks for DVD menus if you own sound sampling software like Adobe Audition or Sony Acid Pro. Both applications ship with music samples and support multiple sound tracks. These applications give you the capability of creating original audio tracks by mixing sound samples. With these applications, you can mix sound samples of guitars, drums, organs, violins, and so on to create a unique soundtrack. If you're creating a soundtrack for a motion menu, create a track that is the same duration as the motion menu background video. Render the soundtrack as a WAV or MP3 file. Import the soundtrack into Adobe Encore DVD and follow the instructions in the previous section to add the audio track to a menu.

Understand Menu Looping

When you create a menu, the default duration is forever and the default end action is stop. However, when you add a video background to create a motion menu, the background video plays and then the default menu end action occurs and the screen turns black. When you create a motion menu, you need to specify the number of times the background video will loop. The number of loops combined with the menu's duration determines the duration for which the menu is displayed.

When you use a video as a menu background, the menu duration is set to the video duration by default. However, when you add additional assets to the menu, such as a background audio track, the menu duration defaults to the last item set. Therefore, if the video and audio tracks are different durations, part of the longer track will be clipped. If your menu duration is longer than your background video duration, the background video will freeze on the final frame until the menu loops. The same holds true with an audio track that is shorter than the menu duration. It won't play again until the menu loops. When you animate buttons, you add another factor to the equation. The button animation plays from the poster frame for the menu's specified duration and then loops back to the poster frame regardless of the duration set for the menu. In other words, if the timeline duration from the poster frame to the end of the timeline is less than the menu duration, the animated button video jumps back to the poster frame as soon as the end of the timeline is reached.

10

Set Motion Menu Looping

When you add a video background or audio background to a menu, the track plays for the specified duration and then the menu end action takes over. The default action for a menu is stop; therefore, the screen turns black unless you specify that the menu loop for a given number of iterations. After you set menu looping, the menu loops for the specified number of times before executing the end action. When you create a motion menu, as a rule, you set the menu to loop forever. To set menu looping, follow these steps:

1. Specify a video background or audio soundtrack for the menu as outlined previously. At this point, you'll still have the Properties palette open.

2. Click the triangle to the right of the Loop# field and then choose an option from the drop-down menu shown next. Alternatively, you can manually enter a number in this field.

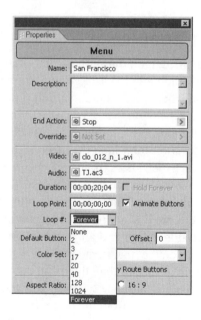

Set Loop Point

When you set menu duration for a motion menu and set the number of loops, you can also set a loop point. When the menu initially loads, the background video plays from frame 1 for the duration specified and then loops. When you set a loop point, the menu loops back to this point, and for this and each consecutive loop, begins playing from the loop point forward. For example, if you set a loop point of 00;00;05;00 (five seconds) when the menu initially loads, menu buttons will be inactive until the background video plays for five seconds. When the menu loops, it loops back to the five-second mark and begins playing. To set a loop point for a motion menu, follow these steps:

1. Select a background video for a movie and specify menu duration and number of loops as outlined previously.

2. In the Loop Point field of the Properties palette, enter a value for the point to which you want the background video to loop. Remember this value is in hours, minutes, seconds, and frames. For example, to specify a loop point of five seconds for an NTSC menu, enter 00;00;05;00 as shown in the following illustration:

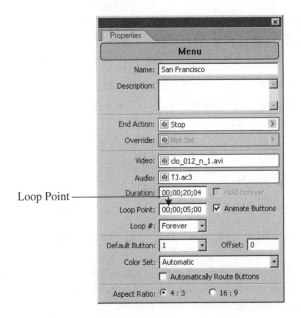

If you set a loop point and have a background audio, the background audio will play from start to finish when the menu is first displayed. Each time thereafter the background audio will start playing at the loop point. For example, if you set the loop point at two seconds, the background audio plays from this point forward after the menu loops one time, and continues looping from that point until the menu is no longer displayed.

Delete Background Audio or Video

When you preview a menu, you may decide that the background audio is not compatible with the background video, or vice versa. You can delete a background audio or video at any time by following these steps:

1. Open the Menus or Project tab.

2. Select the menu that contains the background video or audio you want to delete.

3. Choose Windows | Properties to open the Properties palette.

4. Right-click the Audio or Video field.

5. Choose Delete from the shortcut menu.

Create Animated Buttons

If you've already added a background audio and/or video to a menu, you can take your production to the next level by animating menu buttons. You can also animate buttons that display over a static background. To create animated buttons, follow these steps:

1. Open the Menus tab.

2. Select the menu on which you want to display animated buttons.

3. Choose Window | Properties to open the Properties palette.

4. Click the Animated Buttons check box to activate the feature.

When you animate buttons, they play for the menu duration. If you have added a video background or audio soundtrack to the menu, the duration time has already been set. If the menu has animated buttons over a static background image and no audio background, you'll have to set the duration time as outlined previously in Chapter 8.

Preview Motion Menus

After you select a background video for a motion menu, you can preview the menu to make sure all of your buttons are functioning properly. You can preview an entire project (a task that will be discussed in Chapter 14) or preview a single menu as outlined in Chapter 9. However when you preview a menu, the background video is static, unless you render the menu. To preview a menu and preview the background video at the same time, follow these steps:

1. Open the Menus or Project tab.

2. Select the desired menu, right-click and choose Preview from Here from the shortcut menu. Adobe Encore DVD displays the menu in the Project Preview window.

3. Click the Render Motion Menus button shown in Figure 9-3. Adobe Encore DVD renders the motion menu. As the motion menu is being rendered, the Rendering Motion Menu dialog box appears and a progress bar scrolls, indicating how much of the menu has been rendered.

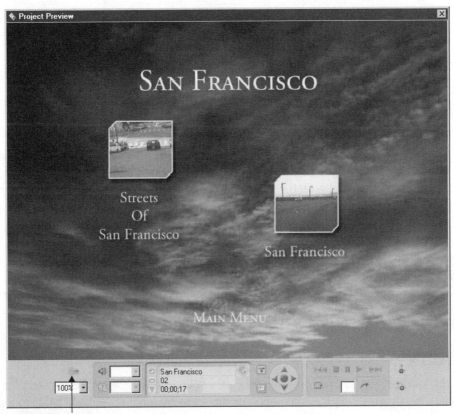

Render Motion Menus

FIGURE 10-3   You can preview motion menus in the Project Preview window.

Render Motion Menus

If you want to preview an entire project that contains multiple motion menus, you must render all motion menus. You can render each individual motion menu while previewing individual project menus as outlined in the previous section, or you can render all motion menus by following these steps:

1. Choose File | Render Motion Menus. After choosing this command, the Render Motion Menus dialog box appears as shown in the following illustration. A blue bar scrolls horizontally to the status of the rendering process. Note that this dialog box monitors the process of each motion menu. After one motion menu is rendered, the dialog box appears again to monitor the status of the next menu build. This process continues until each motion menu in the project is built.

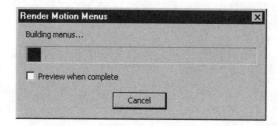

2. Click the Preview when Complete checkbox, and after the last motion menu is rendered, Adobe Encore DVD launches the project in the Project Preview window, starting with the First Play menu or timeline, which gives you the capability to test the entire project.

After rendering the motion menus, you're ready to preview the entire project. However, you will not need to rebuild the motion menus if you exit the application prior to building the project or previewing it in its entirety. When you next open the project, the rendered motion menus are available, as they were saved in the Menus folder of the project folder.

Summary

In this chapter, you learned to spice up your DVD productions by creating motion menus. You learned how to set a motion menu, set menu looping, and add background audio tracks to your productions. You also learned to create animated buttons and specify the duration for which animated buttons play before looping back to the poster frame. In the next chapter, you'll learn how to create custom menus in Adobe Photoshop.

Part IV

Advanced DVD Techniques

Chapter 11

Create Custom Menus in Adobe Photoshop

How To...

- Create a custom menu in Adobe Photoshop
- Understand button set nomenclature
- Dissect a complex button

One of the key features of Adobe Encore DVD is the application's ability to interact with Adobe Photoshop. In previous chapters, you learned to edit a menu in Adobe Photoshop by choosing a menu command while working in Adobe Encore DVD. This is only the tip of the iceberg when it comes to working with Adobe Photoshop. You can use Adobe Photoshop to create compelling, eye-catching menus for your DVD projects from scratch. If you're working on a DVD project for a demanding client or want to create a state-of-the-art DVD to promote your company, your products, or your skill as a videographer, you can use the powerful Adobe Photoshop toolset to create a DVD menu that puts your own stamp of originality on the project.

In this chapter, you'll learn to create a DVD menu in Adobe Photoshop by creating a new document using a preset menu size that conforms to the aspect ratio of your DVD project. You'll learn to work with layers and create layer sets that function as buttons when you import the finished menu into Adobe Encore DVD. You'll learn how to name the layer set so Adobe Encore DVD recognizes it as a button set, as well as name the button set layers so that Adobe Encore DVD recognizes them as subpictures and a placeholder for a video thumbnail. Button sets can have up to three subpicture layers and one layer for the video thumbnail.

Create a Custom Menu in Adobe Photoshop

If you own Adobe Photoshop, you can use the application's toolset to create a custom menu. When you create a custom menu, you can composite images to create a background, as well as create and format text for titles and buttons. You can also create layer sets for use as buttons in your DVD menu. When you create layer sets for use as buttons, you use a naming protocol that enables Adobe Encore DVD to recognize the layer set as a button set.

When you create custom menus in Adobe Photoshop, you're creating a document with square pixels. As you know, video files use rectangular pixels. The aspect ratio of the pixel varies depending on the whether the video is standard 4:3 or widescreen 16:9 aspect ratio. When you create a new document in Adobe Photoshop, you can choose a preset size to suit the aspect ratio and television broadcast standard that matches the DVD project for which you are creating the menu. After you save the menu and import the document into Adobe Encore DVD as a menu asset, the document is converted to rectangular pixels. Table 11-1 shows the television broadcast standards for DVD projects and the preset size you should choose when creating a menu in Adobe Photoshop. Remember, when you create a menu in Adobe Photoshop, you're creating a document with square pixels. When you import the document into Adobe Encore DVD, the document is converted to rectangular pixels for the television broadcast standards shown.

When you create a menu in Adobe Photoshop, you can open bitmap images and drag them into the document you're using to create the menu. When you add images to an Adobe Photoshop document, a separate layer is created for each image. Separate layers are also created for any text or layer sets you create. When you need to create buttons in Adobe Photoshop, you begin by creating a layer set and then renaming the layer set with the proper nomenclature.

To begin the creating a custom menu in Adobe Photoshop, follow these steps:

1. Launch Photoshop and then choose File | New to open the New dialog box.

2. Click the triangle to the right of the Custom field and choose one of the DVD templates shown in the following illustration:

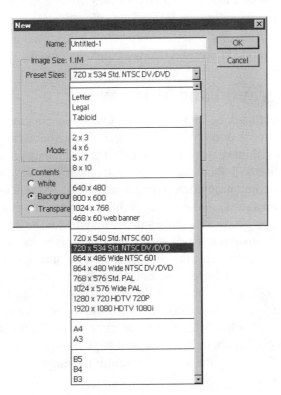

11

DVD Project Television Broadcast Standard	Adobe Photoshop Preset Size
NTSC 4:3	720×534 pixels
NTSC 16:9	864×480 pixels
PAL 4:3	768×576 pixels
PAL 16:9	1024×576 pixels

TABLE 11-1   Adobe Preset Document Sizes for Television Broadcast Standards

3. Accept the default Mode setting of RGB Color.

4. In the Contents field, click a radio button to choose the desired background option.

NOTE   *If you want the menu to have a specific background color, click the background color swatch in the toolbar and choose the desired color before creating the new document. When you create the document, click the Background Color radio button. If you're creating a menu for an NTSC DVD project, remember to keep your color values within a range of 16 to 235. This keeps the colors within the NTSC Safe color palette.*

5. Click OK. Adobe Photoshop opens a blank document.

Create Background and Title Layers

If you're familiar with Adobe Photoshop, you know you can create as many layers as you need to get the job done. When you create a background for your project, you can import images and make a collage or create a gradient background, and then add shapes that you create with the Adobe Photoshop drawing tools. You can also create a striking menu by combining bitmap images and shapes. If you use the Lines tool or create objects with a stroke and transparent fill, remember to specify a line or stroke width of at least three pixels. Anything smaller may flicker on a television screen. Convert any CMYK images to RGB before importing them into a DVD menu.

If you use a gradient to create the background for the menu, use darker saturated colors as lighter colors may have a tendency to look a little washed out on a large television screen. Light colors, however, do work fine for text displayed over a dark background. When you use images as part of a background color correct the images using curves (CTRL-M) or levels (CTRL-L). Use clear, razor-sharp images as they will lose some of their clarity when displayed on a large television screen. If the images are light, consider adjusting the hue and saturation (Images | Adjustments | Hue/ Saturation) of the image.

After you create a background for your DVD menu, it's a good idea to save the file as a menu template before creating layers for text and layer sets for buttons. If your background comprises multiple layers, you can easily edit the layers at a later date by reopening the template.

After you save the file as a template, open the Layer palette, and then choose the Flatten Image command from the Layers pop-up menu. This is optional if you're creating a menu for PAL DVDs but imperative if you're creating a menu for NTSC DVD projects. Adobe Photoshop has a filter you can apply that converts all colors to NTSC safe colors. You could select each layer and apply the filter, but it's easier to apply the command to a flattened image to make sure you don't miss any layers. A single background layer is also easier to deal with in Adobe Encore DVD. In Adobe Encore DVD you can lock the background layer to prevent inadvertently moving it while editing buttons and text. Dealing with one background layer is easier than dealing with multiple image layers, that is unless you intend to convert some of the images on a layer to buttons. In that case flatten all layers with the exception of the ones that house images you intend to convert to buttons. To convert a layer to NTSC safe colors, select the layer and then choose Filter | Video | NTSC Colors. After converting the colors to NTSC, use the Save As command to save the file with a different name than your template.

If you do create vector-based objects such as arrows, circles, rectangles, or lines and you want the capability of moving or otherwise editing these in Adobe Encore DVD, do not flatten these layers with the background layer. Vector-based objects are created using mathematical formulas, therefore you can resize them without losing fidelity. Remember, you can use the Selection or Direct Select tool in Adobe Encore DVD to move and resize objects. You cannot apply the NTSC Colors filter to vector objects. Therefore, when you're choosing colors for a vector object that will be part of an NTSC DVD menu, specify colors using the R, G, and B fields of the Color Picker and enter values between 16 and 235. For example, if you want to create a line with a bright red color, you would enter values of R = 235, G = 16, and B = 16. For a white color enter R = 235, G = 235, and B = 235. If you're choosing colors for a PAL DVD project, you can enter values between 0 and 255 (the full range of values in the RGB color model) in the R, G, and B fields.

After you create the background layer, you use the Text tool to create any text objects for your menu. The Text tool in Adobe Photoshop works just like the Text tool in Adobe Encore DVD. The Adobe Photoshop Character palette is identical to the one you find in Adobe Encore DVD, but with Adobe Photoshop, you can also set many text parameters from the Options bar. Each text object you create appears on its own layer. Don't create any text with a font size smaller than 20 points as it will be hard to read on a television screen. When you specify colors for text objects, you can enter any value between 0 and 255 in the R, G, or B fields for a PAL DVD menu, or values between 16 and 235 if the text is for an NTSC DVD menu.

11

Because text objects appear on separate layers, you can apply layer styles to the text. You can embellish text with layer styles such as drop shadows, outer glow, inner glow, bevel emboss, and so on. Layer styles you apply in Adobe Photoshop are preserved when you import the document into Adobe Encore DVD as a menu. If after you import the menu into Adobe Encore DVD you find that the layer style doesn't suit your DVD project, you can always edit the menu in Adobe Photoshop and remove the layer style or modify it. After you finish creating the required text in Adobe Photoshop, do not flatten the text layers with the background layer; if you do, you won't be able to edit the text when you bring the document into Adobe Encore DVD as a menu.

After you create a document with a background layer and text layers, you have the basis for a main menu in a DVD project. If desired, you can save the document at this point for use as a menu template in Adobe Encore DVD. You can then use buttons from the Adobe Encore DVD Library to flesh out the menu. You can also create buttons within Adobe Photoshop. For each button, you create a layer set and rename each layer using nomenclature that Adobe Encore DVD recognizes as a button set. The upcoming sections show you how to properly name layer sets and how to create graphics for button objects. Figure 11-1 shows a DVD menu under construction in Adobe Photoshop.

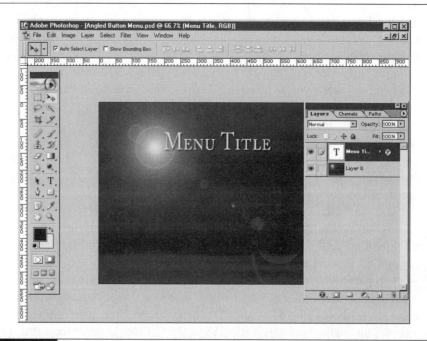

FIGURE 11-1 You can use the Adobe Photoshop toolset to create custom DVD menus.

About Button Sets

If you've explored the menus in the Adobe Encore DVD Library prior to reading this chapter, you know that a button is a layer set (known as a button set in Adobe Encore DVD) that comprises multiple layers. You may have also noticed that the button set, and the layers within the button set, are signified by a unique nomenclature. This nomenclature tells Adobe Encore DVD how to treat each layer within the button set. Button sets can have up to three subpicture color layers and one video thumbnail layer. The following illustration shows a button set as displayed in the Layers palette. The button set is expanded to show three subpicture color layers and the video thumbnail layer.

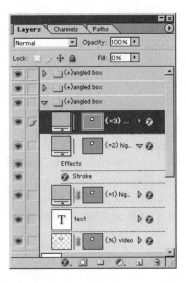

Notice that to the right of each layer—and for that matter the button set—is a character, or characters, in parentheses. The character(s) tell Adobe Encore DVD what function the set or layer performs. Table 11-2 shows the nomenclature you use to signify a button set, as well as the function of each layer of a button set.

Create Button Sets

After you've created your background and text layers, you're ready to add buttons to the menu. When you create a button in Adobe Photoshop, you create a new layer set for each button. After you create a new layer set, you add the desired layers using the nomenclature discussed in the previous section. If you don't use the proper names for a button layer set and each layer in the layer set, Adobe Encore DVD will not recognize the object as a button set.

11

Button Nomenclature	Description
(+)	Designates a button set
(=1)	Designates subpicture color 1
(=2)	Designates subpicture color 2
(=3)	Designates subpicture color 3
(%)	Designates a placeholder for a video thumbnail from the timeline poster frame that is linked to the button

TABLE 11-2   Button Nomenclature

Don't create any buttons that are smaller than 70×60 pixels— a button any smaller will be hard to see on a television screen. It's better to make a button larger than needed because you can always resize them in Adobe Encore DVD. To create a simple button, follow these steps:

1. Open the Layers palette, and then click the Create a New Set button. Adobe Photoshop creates a new set and names it Set 1. Additional sets you create will be named Set 2, Set 3, and so on.

2. Double-click the default set name and rename the set (+). This tells Adobe Encore DVD that the set is a button set. If desired, add a unique name after the closing parenthesis. This will make it easier for you to identify what the button is for when you import the document into Adobe Encore DVD as a menu. For example, you might label a button set (+) **Angled Button**, for a button with an angular shape.

3. Click the Create a New Layer button to create a blank layer with the default name of New Layer.

4. Double click the default layer name and enter (%). This designates the layer is for the video thumbnail.

5. Use the Rectangle tool to create a placeholder for the video thumbnail that has the same proportions as your menu, which of course conforms to the aspect ratio of the DVD project for which you are creating the menu. The easiest way to do this is to create a rectangle the same size as the document. If you use the Rectangle tool and select the Shape Layers option, there's no need to create a layer first as Adobe Photoshop will create a new layer when you create the rectangle. The layer will be a Fill layer with a Vector mask to define the size of the shape. If you use the tool in Fill Pixels mode, you need to create the layer first.

6. Use the Free Transform command (CTRL-T) to resize the thumbnail shape proportionately. After invoking the command, a bounding box with eight handles appears around the shape. Click a corner handle, and drag diagonally while holding the SHIFT key. Alternatively, you can resize the shape by entering values in the Options bar.

7. After you create the shape for the video thumbnail layer, press CTRL-J to duplicate the layer.

8. Rename the new layer (=1). This tells Adobe Encore DVD that the layer is for subpicture color 1. You can further clarify the purpose of this layer by adding text such as Highlight Layer 1. This is not needed by Adobe Encore DVD, but will make it easier to signify what the layer is for when working editing the button set in the Menu Editor. Remember the name that you assign the layer in Adobe Photoshop will be displayed when you open the Layers palette in Adobe Encore DVD.

9. Double-click the layer to open the Blending Options section of the Layer Styles dialog box.

10. In the Advanced Blending section, drag the Fill Opacity slider to 0. If you don't set the Fill Opacity to 0, the button highlight will be the fill, and the video thumbnail will not be visible.

11. Double-click the Stroke style to display the Stroke style options.

12. In the Structure section, enter a value of 3 or 4 in the Size field.

13. Click the triangle to the right of the Position field, and choose Outside from the drop-down list.

14. Click the Color swatch and select a color for the stroke. This will also be the highlight color for the Selected and Activated button states. The highlight appears as a border around the video thumbnail.

That's all there is to creating a simple button. You can place any shape on the video thumbnail layer; however, Adobe Encore DVD will display the video thumbnail as a rectangle that conforms to the DVD project aspect ratio. If you want the video thumbnail to conform the shape you use on the video thumbnail layer, you add a vector or layer mask that conforms to the shape. Adobe Encore DVD uses the shape of the mask to display a portion of the video thumbnail.

After you create the first button, you can duplicate and align the buttons within Adobe Photoshop, or you can save the file with a single button and then duplicate

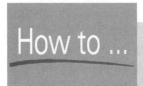 **Create Action and Title Safe Guides in Adobe Photoshop**

When you add items to an Adobe Photoshop document you will be using as a DVD menu, you can reposition the items in Adobe Encore DVD provided you haven't flattened the layers. Or if you prefer, you can create action and title safe guides in Adobe Photoshop by following these steps:

1. Create a new document using the correct preset size for your project.

2. Select the background layer and press CTRL-J to duplicate the layer.

3. Press CTRL-J to duplicate the layer again.

4. Rename the first Duplicate layer *Action Safe* and the second duplicate layer *Title Safe.*

5. In the Layers palette, double-click the Action Safe layer to open the Layer Styles dialog box.

6. In the Advanced Blending section, drag the Fill Opacity slider to 0.

7. Click Stroke to apply a stroke to the layer.

8. Click the triangle to the right of the Position field and choose Center from the drop-down menu. Accept the default value of three pixels for the stroke size, as shown next. If desired, click the Stroke color swatch to choose a different color than the default red.

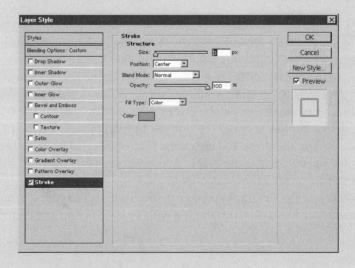

9. With the layer still selected, press CTRL-T to transform the layer. Adobe Photoshop displays the transformation options in the Options bar.

10. Enter a value of 90 in the W and H fields, and then press ENTER to apply the transformation.

11. Select the Action Safe layer in the Layers palette and repeat Steps 6 and 7.

12. Press CTRL-T to transform the layer.

13. Enter a value of 80 in the W and H fields of the Option bar and then press ENTER to apply the transformation.

14. Lock the Action Safe and Title Safe layers by selecting each layer in the Layers palette, and then clicking the Lock icon at the top of the palette.

After performing these steps you'll have two rectangles in the document that conform to the action and title safe areas of a DVD menu, as shown in the following illustration. Place text items within the inner rectangle and important graphic objects within the outer rectangle. Remember that most buttons have text, and as such, should be placed in the title safe area. Hide these layers when you import the menu into Adobe Encore DVD, and the rectangles will not be visible when you build the project.

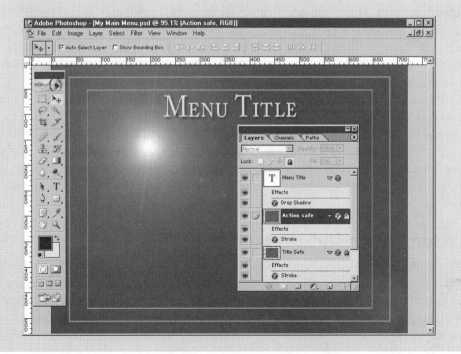

the button as needed after you import the file into Adobe Encore DVD as a menu. You may find the latter method easier. When you import the document as a menu into Adobe Encore DVD, you can duplicate the button as needed. As an added bonus, you can align all of your buttons relative to the safe area.

After you save the document as a PSD file, you use the Import as Menu command to open the file in Adobe Encore DVD. If needed, you use the Edit in Photoshop command to tweak the menu.

Create Subpictures in Photoshop

You can create a three-color subpicture in Adobe Photoshop by creating a button set and then adding three layers. Use the proper nomenclature for each layer as outlined in the "About Button Sets" section of this chapter. Each subpicture layer represents one color of the subpicture. These layers combine to make a single image for the button subpicture. It's important to note that the layers do not represent each button state. If you duplicate layers to create each subpicture layer, the subpicture for each button state is the same image, but different with different colors.

If desired, you can have different shapes for subpicture layer 2 and layer 3. You can place shapes like arrows or lines under the video thumbnail layer. These shapes appear when the button is selected or activated.

When you create a shape that will be the subpicture image, create a simple shape with sharp edges. Do not use gradient fills on the shape. Choose a solid color for each subpicture layer. The DVD specification requires the subpictures be 2-bit graphics. If you use gradients or bitmaps images for a subpicture, you exceed the allowable color palette and your subpicture will not display properly when the DVD is played back. By choosing subpicture colors in Adobe Photoshop, you save time editing the menu color set when you import the file into Adobe Encore DVD as a menu.

When you import the file into Adobe Encore DVD as a menu, a color set is automatically generated based on the colors you selected in Adobe Photoshop. You can use the Menu Default color set in place of the Adobe Encore DVD Automatic color set. You can also edit the default menu color set in Adobe Encore DVD. You'll find detailed information on editing color sets and using the default menu color set in the Chapter 9. You'll also find instructions on how to create a custom color set in Chapter 9.

Create a Sophisticated Button Set

If your project calls for a more sophisticated button shape than your garden variety rectangle, the sky's the limit when you use a layer mask or vector mask to define a video thumbnail layer shape. When you use a layer or vector mask on the video thumbnail layer, Adobe Encore DVD matches the height of the video thumbnail to the height of the layer or vector mask. The shape of the mask determines how much of the underlying video thumbnail is displayed, which gives you a tremendous amount of versatility. If you're creating a DVD for a client who has a uniquely shaped logo, you can create a button that matches this shape. You can also create buttons shaped like rounded rectangles, or you can use Adobe Photoshop's Pen tool to create a freeform shape. As mentioned previously, you should create subpicture shapes that have sharply defined corners. Therefore circles and ovals are not acceptable shapes for buttons when you are going to duplicate the shape and use it as the basis for your subpicture layers. If you use an oval or circle as a shape for a button's video thumbnail layer, use a different shape such as a line under the circle or oval for the subpicture layers. If you use the rounded rectangle tool as the basis for a button shape, do not exceed a radius of five pixels. To create a uniquely shaped button using a layer mask, follow these steps:

1. In the Layers palette, click the Create New Layer set button.

2. Rename the layer set (+), as outlined previously.

3. Create a new layer.

4. Rename the new layer (%). This is your video thumbnail layer.

5. Select the desired shape tool. In the Options bar, click the Fill Pixels button. If you want the shape to be a specific size, you can click the Geometry Options icon in the Options bar; then click the Fixed Size radio button and enter the desired size in the W and H fields.

6. Create the desired shape within the document. You can use any solid color for the shape, as it won't show up in your DVD menu. Black is always a good choice, unless of course your menu background is black or another dark color. If this is the case, choose a contrasting color so you'll easily be able to see the shape.

7. Select the Magic Wand tool and click inside the shape to create a selection. The tool will select the entire shape as you're using a solid color.

8. Choose Layer | Create Layer | From Current Selection. Adobe Photoshop creates a mask layer.

9. In the Layers palette, click the Add a Layer Style button and choose Stroke to open the Stroke section of the Layer Style dialog box.

10. Click the triangle to the right of the Position field and choose Inside from the drop-down menu, as shown next. This creates a frame around your video thumbnail. The frame is optional but adds a finishing touch to the button.

11

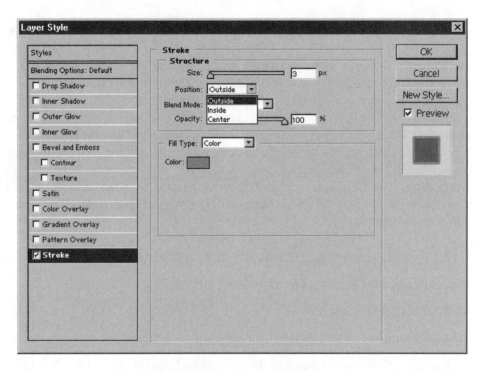

11. Click the color swatch and choose a color for the stroke from the color picker. If you're creating a button for an NTSC DVD menu, choose color values from 16 to 235.

12. Select the layer in the Layers palette and press CTRL-J to duplicate the layer.

13. Rename the duplicated layer to (=1). If desired, add additional text after the closing parenthesis to help you identify the layer after you import the menu into Adobe Encore DVD.

14. In the Layers palette, double click the Layers style icon to the right of the layer's name. Adobe Photoshop opens the Blending Options section of the Layer Styles dialog box.

15. In the Advanced Opacity section, drag the Fill Opacity slider to 0, as shown next.

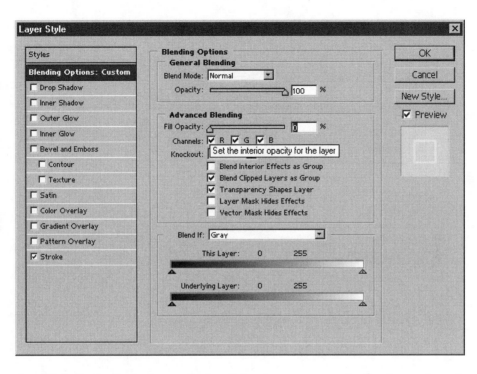

16. Click Stroke in the Layer Styles dialog box to access the Stroke section of the Layer Styles dialog box.

17. Click the triangle to the right of the Position field and choose Outside from the drop-down menu. If you accepted the Inside option from the original layer, the button frame will disappear whenever the button is highlighted.

18. Click the color swatch and choose the desired highlight color from the Color Picker. If you're creating a button for an NTSC DVD menu, remember to specify color values within a range from 16 to 235. The following illustration shows the Stroke section of the Layer Styles dialog box after the stroke for the (=1) layer has been modified. A bright red color has been selected for the highlight color.

11

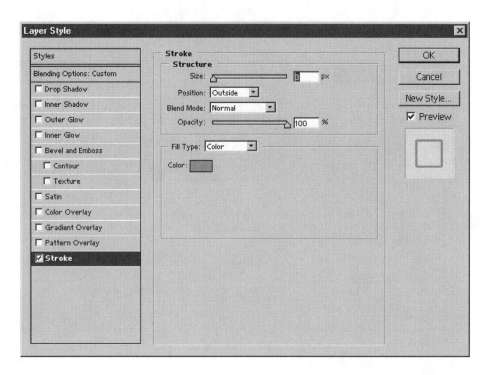

If desired, you can create additional subpicture layers as outlined previously. With a bit of work in Adobe Photoshop and layer masks, you can create interesting button shapes for your DVD projects. With a bit of experimentation and the powerful Adobe Photoshop toolset, you can create unique menus and buttons that make your finished DVDs stand out in a crowd. Figure 11-2 shows a finished button in Adobe Photoshop. The Layers palette is included in this screenshot to show you the various components of the button set with title text added as a separate layer.

You can also create a custom button shape by using a vector layer mask. The steps are similar, but instead of creating a mask by creating a selection, you create a vector path over an object you create with one of the drawing tools. The vector path is the same shape as the button you want to create. You then create a vector mask by choosing Layer | Add Vector Mask | Current Path.

Analyze a Button in Adobe Photoshop

If you've used any of the buttons from the Goodies folder on the Adobe Encore DVD installation disc, you know there are some uniquely shaped buttons from which

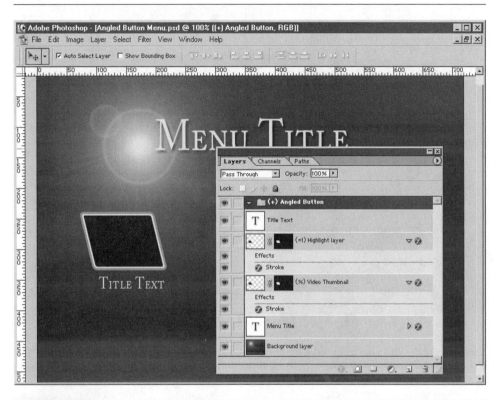

FIGURE 11-2   You can create a custom button using a layer mask.

to choose. When you want to find out what makes a button tick and then use the knowledge to create similar buttons, add the button you want to examine to a blank menu, select the menu, and then choose Menu | Edit in Photoshop. After the menu opens in Adobe Photoshop, select the button set, and then open the Layers palette. The following illustration shows a button from the Goodies folder in a blank 4:3 NTSC menu with a black background. The menu was opened in Adobe Photoshop with all button set layers displayed in the Layers palette as shown next. The Layer effects have been expanded to illustrate the effects applied to each layer.

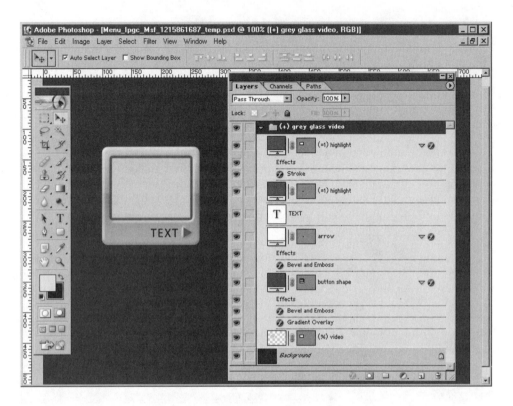

When you analyze a button in Adobe Photoshop, the first thing to do is open the Layers palette. From within the Layers palette, you'll be able to isolate each layer of the button set to see what shapes were used to create it and what Layer styles were used. Momentarily hide the other layers by clicking the eyeball icon in the layers Show/Hide column. After analyzing the button set in Adobe Photoshop, you can take what you've learned and use the knowledge to create similar buttons for your custom menus.

Summary

In this chapter, you learned to create custom DVD menus in Adobe Photoshop. You learned to create button sets and subpicture layers. You also learned to create unique button shapes by using layer masks or vector masks, and how to analyze button sets in Adobe Photoshop. In the next chapter, you'll learn to work with alternate audio tracks.

Chapter 12

Advanced DVD Techniques

How To...

- Set default audio track
- Add alternate audio tracks
- Create alternate audio track menu
- Manage alternate tracks

Audio is a fact of life for just about every DVD project you'll create. Depending on your needs, your organization's needs, or your client's needs, you may have to create a project with multiple audio tracks. When you create a project with multiple audio tracks, you give your viewing audience a choice. For example, you can create a DVD disc with alternate tracks that have comments from the director or creator of the movie. Another use for alternate audio tracks is when your viewing audiences speak different languages.

In this chapter, you'll learn to add alternate audio tracks to a video timeline. You'll also learn to manage alternate audio tracks, as well as set up a menu for selecting alternate audio tracks.

Understand Alternate Audio Tracks

Set top DVD players can play only one soundtrack at a time. If you need to add sound effects or special effects to a DVD production, you will need to mix these sounds in a sound-editing application, such as Adobe Audition or Sony Sound Forge, or within a video-editing application that supports multiple sound tracks. After you mix the effects with the soundtrack, export the soundtrack in a format recognized by Adobe Encore DVD and then import the audio track into your project and add the track to the desired video timeline.

When you create a DVD project that uses multiple audio tracks, Track 1 is used for the video soundtrack. If you import your audio and video assets separately, Track 1 should be used for the default video soundtrack. You can use additional audio tracks for soundtracks in foreign languages, director's comments, soundtracks using Dolby 5.1 Surround Sound, and so on.

Set Default Audio Track

When you create a DVD project with multiple audio tracks for each timeline, you can set the default audio track for the entire project. When you specify a default audio track for a project, a timeline will default to that audio track if the audio track specified by a menu button does not exist. For example, if several of the timelines in your

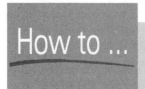 **Create Alternate Audio Tracks**

If you're part of a small production company, you're probably wearing more than one hat. In addition to authoring DVD projects, you may also be responsible for editing audio and video clips for the project. When you have a project that employs alternate audio tracks, it's useful to treat audio and video as separate entities. If your video-editing software supports exporting audio and video as separate tracks, you're all set.

If your video-editing application supports multiple audio tracks, you can add a track for Director's comments. You can either record the track using a microphone connected to your PC or insert a prerecorded soundtrack. You can also create alternate tracks for different languages. Create one track for the music soundtrack and another for the language. The following illustration shows a timeline from a video project with multiple soundtracks being created in Adobe Premiere Pro:

From within your video-editing application, you can save different versions of the project, one for each alternate track. Render the video only, and then render each alternate audio track as audio only. Render your audio and video in DVD legal formats supported by Adobe Encore DVD. After that, it's a simple matter of creating a timeline for your video and then adding each alternate track. Your video will be perfectly synchronized with each audio track.

project have an audio track with director's comment and one timeline does not, you can program a button to play the director's comments track when activated; however, when the timeline with no director's comments track is selected, the default audio track is played. To set the default audio track for a project, follow these steps:

1. Open the Disc tab.

2. Choose Window | Properties to open the Properties palette.

3. Click the triangle to the right of the Set Audio field and choose the desired audio track from the drop-down list, as shown next. Note that if you use the default setting of No Change, the audio track as specified in the DVD player becomes the default audio track.

Add Alternate Audio Tracks to the Timeline

When you create a project that requires multiple audio tracks, you import your assets as with any project. Before you build your navigation menus, you create timelines for your video assets. If your video assets have embedded audio, Adobe Encore DVD uses Audio Track 1 for the embedded audio. You can add up to seven more audio tracks to a timeline. If you've modified Preferences to display more than one audio track, you'll have blank audio tracks to which you can add audio. Figure 12-1 shows a timeline with multiple audio tracks. To add alternate tracks to a timeline, follow these steps:

1. Open the Timelines tab.

2. Double-click the desired timeline to open it in the Timeline window.

3. Click the Project tab and then resize the window so you have access to both the Timeline and Project tabs.

4. Create a new audio track by doing one of the following:

■ If you've specified more than one audio track, drag an audio asset from the Project tab to the applicable audio track.

■ Right-click the audio track, choose Add Audio Track from the shortcut menu to create a blank audio track, and then drag and drop the desired audio asset to the new track.

■ Drag an audio asset from the Project tab and drop it into the Timeline window. Adobe Encore DVD creates a new audio track.

5. If applicable, click the triangle to the right of the Language field and choose the desired language code from the pop-up list. If you choose a language option, when the DVD is played and the audio track is selected, the language is momentarily displayed on screen.

After you add alternate audio tracks to a timeline, it's a good idea to play them to make sure they are properly synchronized with the video. When you add alternative clips to a timeline, they snap to the beginning of the timeline. You cannot nudge or otherwise move the clip to change the point at which the clip begins to play. If, however, a clip is not playing as desired, or the audio clip has other problems, such as excessive volume, you can rectify these problems by editing the audio asset in an audio-editing application.

12

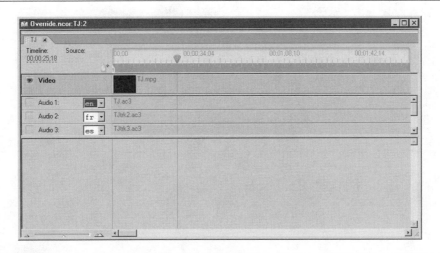

FIGURE 12-1   You can add up to eight audio tracks to a timeline.

Edit Soundtracks

If you use multiple soundtracks in a DVD project, you may find that some of the tracks need editing. For example, one or more tracks may be louder than others. If you miss details like this, your DVD project will be tagged as unprofessional. If you own audio-editing software such as Adobe Audition or Sony Sound Forge, you can edit an audio clip. When you edit an audio clip you can correct for any deficiencies such as hiss, excessive volume, noise and so on. Depending on the application you use to edit your audio clips, you may be able to view the video track while editing the audio track. This option is handy if you need to add narration to an audio clip that needs to be synchronized with an event in your video clip. The following illustration shows Adobe Audition in multitrack mode with the video and audio tracks for an MPEG 2 movie displayed. The video for the clip is displayed in a separate monitor window. With a setup like this, you can easily add additional audio tracks such as sound effects, or use a microphone attached to your PC to create a voice-over that is perfectly synchronized with the video.

When you're editing an audio clip, you can use filters to normalize an audio track, adjust the volume, add special effects such as reverb, echo, and so on. When you open an audio clip in an audio-editing application, you can analyze the waveform. If you have an audio clip with exceptionally high spikes, this indicates the possibility of excessive volume, which can lead to distorted sound. You can use the tools and menu commands in your audio-editing application to correct problems like this. The following illustration shows an audio track as viewed in Adobe Audition:

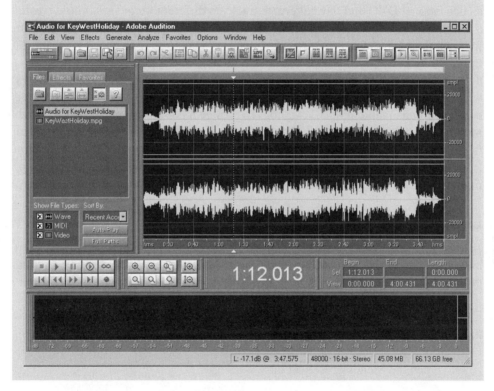

Remove Alternate Audio Tracks from a Timeline

If you decide an alternate audio track is no longer needed in a timeline, you can delete it at any time. To delete an alternate audio track, follow these steps:

1. Open the Timelines tab.

2. Double-click the desired timeline to open it in the Timeline window.

3. Right-click the audio track you want to remove from the timeline to display the shortcut menu.

4. Choose Remove Audio Track.

Create a Menu for Alternate Audio Tracks

When you have multiple audio tracks, you must create a method for your DVD viewers to select the audio track they want to hear. For example, if you've created an educational DVD and have alternate tracks in different languages, you'd create a menu that enables viewers to select the desired language. You can create a menu for your alternate tracks using text items that you've converted to buttons. For more information on converting text objects to buttons, see Chapter 9. After you create the text button, you create a link for the button and specify which audio track is played when the button is clicked. When you create the button, it's a good idea to give the button a unique name like "5.1 Audio." To change a button's name, select it in the Menus tab, open the Properties palette, and then enter the desired name in the Name field. Giving a button a unique name will make it easier to identify what the button is for when you open the Menus tab and select the menu on which the button appears. Figure 12-2 shows a typical menu for selecting audio tracks.

Link to Alternate Audio Tracks

After you create a menu with a button for each alternate audio track, you link each button to another menu in your project. The actual menu will vary from project to project, but as a rule, you'll link back to the title or main menu in the project. When you create the button link, you also specify which audio track plays when the button is clicked. To specify which audio track plays when a button is clicked, follow these steps:

1. Open the Menus or Project tab.

2. Double-click the menu you created for audio choices.

3. Select the desired button.

4. Choose Window | Properties to open the Properties palette.

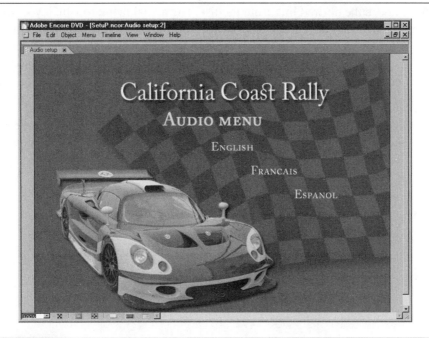

FIGURE 12-2   You can create a menu for selecting alternate audio tracks.

5. Click the arrow to the right of the Link field, and from the drop-down list, choose Specify Other to open the Specify Link dialog box, shown next:

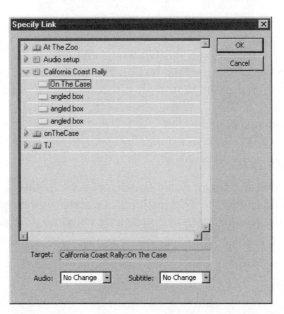

6. Choose the desired link from the list. The actual menu to which you set the link will vary depending on your project. In many cases, you'll link back to the main menu after the viewer makes a selection for audio. However, if you have other set up options such as subtitle tracks, you would link back to the default button of the setup menu that includes both audio and subtitle choices. After choosing a link, Adobe Encore DVD displays it in the target field.

7. Click the triangle to the right of the Audio field and choose the desired audio track from the drop-down list. The following illustration shows the Specify Link dialog box as it appears after choosing a link and specifying an audio track.

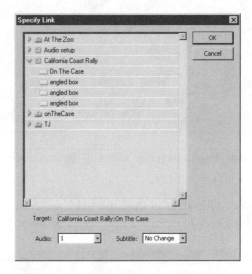

The audio track that you specify will be the audio track that is played for all timelines in the project. For example, if track 2 of each timeline contains director's comments, you create a menu button for director's comments and link to track number 2 as outlined in Step 7. When viewers click this button, every timeline that they select will play track 2.

Manage Alternate Tracks

When you have several alternate audio tracks for each timeline, it's imperative you organize everything so that the right track plays when a viewer makes a selection. Your first line of defense for organizing your tracks is naming the audio tracks after you import them as assets. Choose logical names that reflect the content of the audio clip; for example, "French language track." After you have your tracks named properly, the next step is to add them to the proper tracks in the associated video timeline.

You should specify the first audio track as the default audio track. When you add additional audio tracks to the timeline, make sure you use the same track number for a particular audio selection. For example, you may want to use track 2 for the audio track with director's comments. If you do so, make sure that you add the director's comments audio clip to audio track 2 for every timeline in your project. If you're working with a large number of timelines, you may want to put pen to paper to plan out your timelines ahead of time. Another handy tool in this regard is creating a table in your word-processing application.

If you take a few minutes to name and organize your audio assets before adding them to audio tracks, you'll save a lot of time at the end of the project when trying to figure out why the Spanish language soundtrack plays in track 2 of scene 2 instead of the director's comments. Remember, when you preview a project, you can toggle through each timeline audio track. For more information on previewing projects, refer to Chapter 14.

Create Audio-Only Timelines

If you decide your project would benefit from some special effects of the audio variety, you can create audio-only timelines. You can use audio-only timelines to introduce your DVD presentation with a musical fanfare, or you can create special effects as outlined in the upcoming "Create Special Effects with Audio Timelines" section. To create an audio only timeline, follow these steps:

1. Choose File | Import as Asset to open the Import as Asset dialog box.

2. Select the desired audio file and click Open.

3. Click the Project tab.

4. Select the desired audio clip and then click the Create New Timeline button. Adobe Encore DVD creates a timeline with audio and no video.

12

NOTE   *You can set an audio-only timeline as First Play. However, the customary First Play icon does not appear next to the timeline's title in the Project or Timelines tab. In spite of this, the timeline will play first after the DVD is loaded into a set top DVD player.*

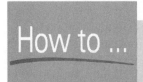 **Create Special Effects with Audio Timelines**

When you think of a DVD presentation, you think of video clips like movies and documentaries. While sound is an important part of the equation, the video is most often the highlight. However, there will be situations when you can use audio-only timelines to augment a DVD presentation. For example, you can create audio-only timelines that contain a brief introduction to each scene. Link the menu button for the scene to the audio-only timeline, and then use the end action for the audio-only timeline as a link to the scene timeline.

You could also use an audio-only timeline as a transition between a scene and the menu. For example, if you have three navigation menus in your DVD project, set the end action for each timeline in a menu as a link to an audio-only timeline. Set the end action for the audio-only timeline as a link back to the default menu button. Of course, the content of the audio track is left up to your discretion.

You can also let audio be the star of your presentation. For example, if you're creating a DVD project that features the works of a symphonic orchestra, the audio will be the star of the presentation. In this regard, you'll want to use PCM audio for your tracks, as this will give you the best sound quality—though it does use the most disc space. If you edit the video for the presentation, intersperse the video of the orchestra playing with some video that augments the audio. For example, if the orchestra is playing the "Spring Allegro" section of Vivaldi's *Four Seasons,* add some video clips of clouds over bucolic scenes of trees, birds, babbling brooks, and so on.

Summary

In this chapter, you learned to work with alternate audio tracks. You learned to add alternate audio tracks to timelines, create menus for alternate audio tracks, create menu links for alternate audio tracks and create audio only timelines for use as special effects. You also learned how to manage alternate audio tracks so that the proper clips are associated with the same audio track in each timeline. In the next chapter, you'll learn to add subtitles to your productions.

Chapter 13

Add Subtitles to a DVD

How To...

- Add subtitles
- Set subtitle duration
- Set subtitle character attributes
- Create subtitles

When you create a DVD project, you can have up to 32 subtitle tracks per video timeline. Subtitle tracks give you the capability to augment a DVD presentation with written text that overlays the video track. Subtitle tracks can be used to present additional information or as an aid for viewers with impaired hearing. When you create subtitle tracks in Adobe Encore DVD, you can specify the font used to display the subtitle, the font color, and so on.

In this chapter, you'll learn to add subtitle tracks to your DVD projects. You'll learn to create subtitles by importing external files and how to create subtitles manually in Adobe Encore DVD. You'll also learn to create menus that viewers use to choose subtitle tracks, as well as how to specify color sets for subtitle tracks.

Understand Subtitles

You can have up to 32 subtitle tracks in a project. Each subtitle track is comprised of subtitle clips that overlay the video for which the subtitles are created. Each subtitle is added as a track to a video timeline. After adding a subtitle track, you can import subtitles or manually create them. Whether you import or manually create subtitles, they do not appear as a continuous track but appear as individual subtitle clips along the subtitle track. Figure 13-1 shows a video timeline with three subtitle tracks.

After you build a DVD project, each subtitle clip appears as an overlay on top of the video track. A set top DVD player can display only one subtitle track at a time. When subtitle text is displayed, it is displayed with a three-color timeline color set, similar to the color sets used to display subpictures. A timeline color set is applied to the entire subtitle track. You can use the default timeline color set to display subtitles or you can specify the desired colors for a timeline color set. The three colors define the fill color, the stroke (or text border) color, and the anti-alias color (the color used to blend the subtitle text with the video) for subtitle text. When you specify colors

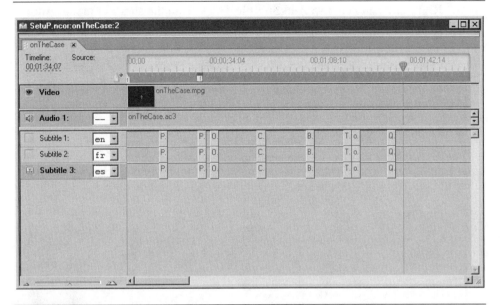

FIGURE 13-1   You can add up to 32 subtitles to a video track.

for a timeline color set, you can also specify the opacity for each color. Timeline color sets are discussed in detail in the upcoming Define Subtitle Color Set section.

Add Subtitle Tracks to a Timeline

When you decide to augment a video track with text subtitles, you add the subtitle track to the desired timeline. You can add up to 32 subtitle tracks to a video timeline. To add a subtitle track, follow these steps:

1. Open the Timelines tab.

2. Double-click the desired timeline to display the timeline in the Timeline window. If you have already specified the subtitle preferences to determine how many subtitle tracks each project's timeline will have, you will have that many subtitle tracks with which to work. If you have not specified a default number of subtitle tracks, you can add up to 32 tracks. You can also add a track when you have used the default number of subtitle tracks (provided the default number of subtitle tracks is less than 32) and need to add more.

13

3. Right-click and choose Add Subtitle Track from the shortcut menu. Adobe Encore DVD adds a subtitle track, as shown next:

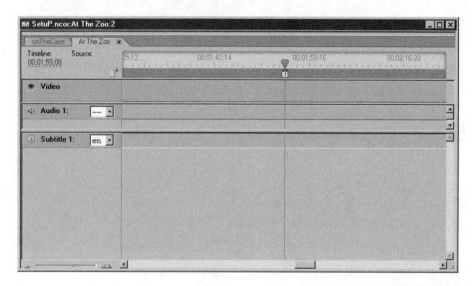

4. If applicable, click the triangle to the right of the Language field and select the desired language from the drop-down list. If you select a language, when viewers play the DVD and select the subtitle track, the language is momentarily displayed on screen.

After you add one or more subtitle tracks to a video timeline, you're ready to add the subtitle text. You can add the subtitle text either by importing a file or by manually creating the text in the Timeline Monitor window.

Set Default Subtitle Track

When you have a project with several subtitle tracks, you can set the default subtitle track, which is the track that will be played by the DVD player unless a viewer makes a different choice. To set the default subtitle track for a project, follow these steps:

1. Choose Window | Disc to access the Disc tab.

2. Choose Window | Properties to open the Properties palette.

3. Click the triangle to the right of the Set Subtitle field, and choose the desired track from the drop-down list. Note that the default setting of No Change uses the subtitle track as specified in the DVD player. Also note that you can also choose Off to play no subtitle track until the viewer chooses an option from the DVD disc setup menu you create.

Import Subtitle Files

You can import files for use as subtitles. Adobe Encore DVD supports FAB images and Captions Inc. scripts. As these are image files, you cannot reposition or edit the files. To import a Captions Inc. or FAB Images script to a subtitle track, follow these steps:

1. Create a subtitle track as outlined previously.

2. Choose Timelines | Import Subtitle Track, and then select Captions Inc. Script or FAB Images Script from the drop-down menu to display the Open dialog box. Alternatively, you can right-click within the Timeline window, choose Import Subtitles, and then choose the desired format from the shortcut menu.

3. Choose the desired file and then click Open. If you've selected the FAB Images Script option, select the folder that contains the Image files and then click OK to access the Map Color Options dialog box; then proceed to Step 4. If you've chosen Captions Inc., the Import Subtitles dialog box appears; proceed to Step 6.

4. Select an eyedropper, position it over the desired area in the image, and then click to sample the color. In all, you have three eyedroppers, which you use to map the image color to the subtitle color set. The eyedroppers are as follows:

    ■ **Background**   Maps the image background color to the timeline color set. Click the eyedropper over the image background to sample the color.

    ■ **Fill/Color 1**   Maps the image text color to the timeline color set. Click the eyedropper over the body of the text to sample the color.

    ■ **Outline/Color 2**   Maps the image text outline color to the timeline color set. Click the eyedropper over the outline of the text to sample the color.

5. Click OK to open the Import Subtitles dialog box.

6. Specify the following settings:

    ■ In the Subtitles section, click the triangle to the right of the tracks menu and choose the desired track from the drop-down list. Subtitle Preferences will govern the number of tracks you have available unless you have manually added subtitle tracks to a project. In that case, the drop-down list shows the number of tracks you've added to the project. If you have not added subtitle tracks to the project, or have used the default number of subtitle tracks as specified in Subtitle Preferences, the only available option is New.

13

- If desired, in the Subtitles section, click the triangle to the right of the Language field and choose a language from the drop-down list, otherwise the language defaults to the language specified in Subtitles preferences.

- In the Color Set section, click the triangle to the right of the Color Set field and choose an option from the drop-down list.

- In the Color Set section, click the triangle to the right of the Group field and choose which group of colors from the color set to apply to the imported subtitles. Your choices are Group 1, Group 2, and Group 3.

- In the Timecode section, choose Relative or Absolute. Choose Relative if you have trimmed the timeline video clip in Encore DVD. This opens a text field that enables you to enter the timecode at which you want the subtitles to appear. For example, if you trim five seconds off a clip, you would enter a value of five seconds in this field to properly synchronize the subtitles with the trimmed clip. If you choose Absolute, the subtitles are imported using the timecodes specified in the subtitle file you are importing.

7. Click OK to import the subtitles.

 To create your own Captions Inc. or FAB Images subtitles, you'll need the proper software. One such application is Title Factory. You can download a trial version of Title Factory at www.titlefactory.com.

Import Subtitle Text Files

You can also import subtitle text scripts. A subtitle text script specifies the timecodes at which a subtitle begins and ends, as well as the subtitle text. When you import a subtitle script into a timeline, you can specify font attributes for the text, as well as which color group is used to display subtitle text. To import a subtitle script, follow these steps:

1. Create a subtitle track as outlined previously.

2. Choose Timelines | Import Subtitles | Text Script to access the Open dialog box. Alternatively, you can right click the desired subtitle track and choose Import Subtitles | Text Script from the shortcut menu.

3. Select the desired subtitle script and click Open. Adobe Encore DVD displays the Import Subtitles dialog box, shown next. Notice that the Title Safe option is enabled, which gives you a visual reference when moving subtitle text either manually or by entering values. Also notice the text box that surrounds the subtitle text. You can move the text box by clicking and dragging the center handle; resize the text box by clicking and dragging any of the perimeter handles.

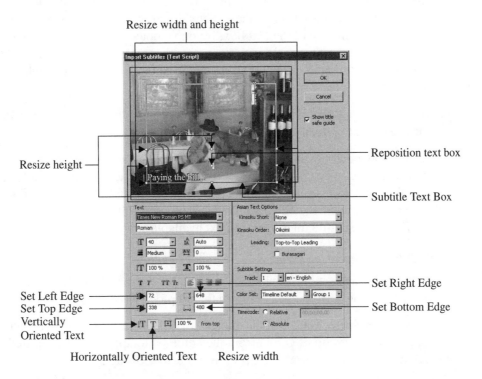

4. In the Text area, specify the character attributes for the text. These are the same options you have available in the Character palette previously discussed in Chapter 8.

5. In the text box alignment area, accept the default values or enter values to set the top and edges of the text box. When you enter different values, the text box is resized accordingly. Make sure you do not enter values that cause the text box to exceed the boundaries of the title safe area, which is possible when you manually enter values. When you use the handles to move or resize the text box, you cannot exceed the boundary of the title safe area; in fact, Adobe Encore DVD will snap the text box to the applicable border if you release the mouse button when the text box boundary is beyond the title safe area.

6. Accept the default text orientation option (horizontally oriented text) or click the Vertically Oriented Text button. If you decide to orient the text vertically, you will have to resize the text box.

7. Enter a value in the text field to the right of the text orientation buttons. If your text is oriented horizontally (the default), the value specifies the percentage that the text appears from the top of the text box. The default value of 100 (percent) places the text at the bottom of the text box. Enter a lower value to move the text towards the top of the text box; for example, a value of 50 places the text in the middle of the text box. If you've opted for vertically oriented text, this value determines the percentage that the text is placed from the right side of the text box. A value of 0 places the text on the right border, while a value of 100 places the text on the left border.

8. If your project subtitles are in Asian text, choose the appropriate settings in the Asian Text options area.

9. In the Subtitle settings area, accept the default subtitle track, which is the track you created or selected prior to importing the subtitle text. If the project has no subtitle tracks, or you have used up the number of subtitle tracks specified in Subtitle Preferences, the only option is New, in which case Adobe Encore DVD creates subtitle track 1 or the next available subtitle track.

10. If desired, select a language from the Language drop-down list.

11. In the Color Set section, click the triangle to the right of the Color Set field and choose the desired color set from the drop-down list.

12. In the Color Set section, click the triangle to the right of the Group field and choose which group of colors from the color set to apply to the imported subtitles. Your choices are Group 1, Group 2, and Group 3.

13. In the Timecode section, choose Relative or Absolute. Choose Relative if you have trimmed the clip in Encore DVD. This opens a text field that enables you to enter the timecode at which you want the subtitles to appear. For example, if you trimmed five seconds off a clip, you would enter a value of five seconds in this field to properly synchronize the subtitles with the trimmed clip. If you choose Absolute, the subtitles are imported using the timecodes specified in the text file you are importing.

14. Click OK to import the subtitle text script. Figure 13-2 shows the Timeline Window after importing a subtitle text script.

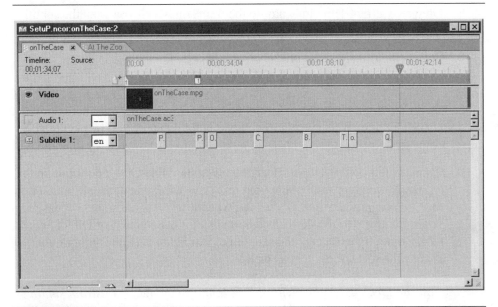

FIGURE 13-2   You can import subtitle scripts to a subtitle track.

In Figure 13-2, notice that each subtitle clip appears as a block on the timeline track, which represents the timecode at which the subtitle clip is displayed and the display duration.

Create Subtitles in Encore

You can create your own subtitles by using the Text tool to type text directly into the Monitor window. When you create subtitles in this manner, you set the frame on which the subtitle begins, and also set the text attributes. To create subtitle text for a timeline, follow these steps:

1. Open the Timelines tab.

2. Double-click the desired timeline. Adobe Encore DVD displays the timeline in the Timeline window, and the first frame of the timeline is displayed in the Monitor window.

3. Add a subtitle track as outlined previously.

4. In the Monitor window, click the Title Safe button. Note that if you have not added a subtitle track to the timeline, you will not be able to display the Title Safe area in the Monitor window.

13

5. Drag the Current Time Indicator to the frame on which you want the subtitle text to appear.

6. Select either the Vertical or Horizontal Text tool.

7. Choose Window | Character to open the Character palette.

8. Select the desired font face, set the font size, and so on. Note that you will not be able to set font color as subtitle text color is determined by the timeline color set.

9. Click the left border of the title safe area in the Monitor window and begin typing. As you type, make sure your text stays within the title safe area. Press ENTER to begin a new line. If after you type the text it is not aligned as desired, you can move it to the desired position using the Selection or Direct Select tool. Do not try to change the size of the text box with either tool as you'll change the scale of the text as well.

TIP *You can create a text box within the Monitor window to constrain subtitle text to the title safe area. Select the Text tool, click the left corner of the title safe area, drag diagonally beyond the right boundary of the title safe area, and then release the mouse button. Adobe Encore DVD snaps the text box to the right boundary of the title safe area. When you enter text and reach the end of the text box, Adobe Encore DVD wraps the text to a new line.*

10. To add additional subtitles to the timeline, repeat Steps 4 through 8.

Define Subtitle Color Set

When you create subtitles, Adobe Encore DVD uses the default timeline color set to determine the fill, outline, and antialias color for subtitle text. Each timeline color set has three color groups, from which you can choose when defining subtitle display properties. You can modify the timeline default color set by specifying different colors and opacities for each group. You can also create a custom color set. To define a subtitle color set, follow these steps:

1. Choose Edit | Color Set | Timeline to open the Timeline Color Set window shown in Figure 13-3.

2. To create a new color set, click the New Color Set button to open the New Color Set name dialog box.

3. Enter a name for the color set and click OK. The Timeline Color Set dialog box refreshes to show the name of the new color set in the Color Set field. The new color set inherits the colors from the color set you selected prior to creating a new color set. The following steps show how to change individual colors for each group.

4. Double-click a group color that you want to define. Adobe Encore DVD displays the Color Picker window, shown next:

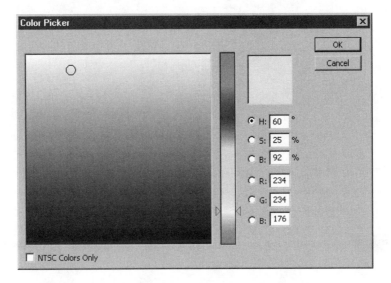

5. Click the NTSC Colors Only to limit your selections to colors in the NTSC Safe color palette.

6. Specify the desired color by dragging the color spectrum sliders and then clicking inside the color well. Alternatively, you can enter values in either the H, S, B fields or the R, G, B fields. Note that if you've enabled NTSC safe colors, Adobe Encore DVD automatically replaces any non-NTSC safe color values with values that create the closest match to an NTSC safe color. The current color appears in the top of the Color Preview window, while the selected color appears in the bottom half of the window.

7. Click OK to set the color.

8. Click the triangle to the right of the Opacity field and choose the desired value from the drop-down list.

13

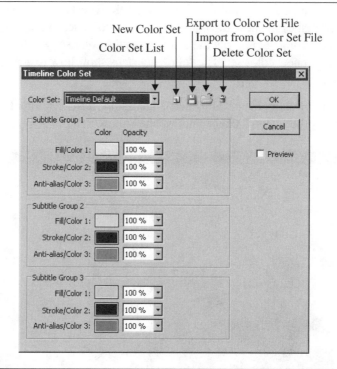

New Color Set
Export to Color Set File
Color Set List
Import from Color Set File
Delete Color Set

FIGURE 13-3   You can define subtitle text colors by creating timeline color sets.

9. Repeat Steps 4 through 8 to set other colors.

10. Click OK when you've defined the desired colors.

You can create as many color sets as needed for your project. When you save the DVD project, the color sets are saved in the project folder.

TIP   *You can modify the default timeline color set by selecting it from the Color Set dropdown list and then following Steps 4 through 10 listed previously.*

Export and Import Timeline Color Sets

After you create a timeline color set or modify the default timeline color set, you can export the color set for use in other projects. When you export a color set, you export a file with the CS extension, which is the same extension used for menu color sets. Therefore, you are advised to give each timeline color set a unique name that readily identify them as timeline color sets. To export a timeline color set, follow these steps:

1. Choose Edit | Color Sets | Timeline to display the Timeline Color Set dialog box.

2. Click the triangle to the right of the Color Set field, and choose the desired color set from the drop-down list.

3. Click the Export to Color Set File button shown previously in Figure 13-3. Adobe Encore DVD displays the Save Color Set dialog box.

4. Enter a name for the color set; then navigate to the folder in which you want to save the file and click Save.

If you save a lot of timeline color sets and menu color sets, it's a good idea to set up separate folders for each color set type on your system. When you name color sets, give them descriptive names such as *gold_yellow_bronze_tl*. Listing the colors makes it easier for you to ascertain if that particular color set is suited for your project. Adding *tl* to the end of the color set name identifies it as a timeline color set.

After exporting a color set, you can import it into any project. To import a previously saved timeline color set, follow these steps:

1. Choose Edit | Color Sets | Timeline to display the Timeline Color Set dialog box.

2. Click the Import from Color Set File button shown previously in Figure 13-3. Adobe Encore DVD displays the Import Color Set File dialog box.

3. Select the desired color set file and click Open.

When you import a color set into a project, it is saved in the project folder.

Specify a Timeline Color Set

When you create a project that uses subtitles, Adobe Encore DVD automatically sets the color set for each timeline. When you create additional color sets for a project or import color sets, you can specify which color set is used to define a timeline's subtitle text by following these steps:

1. Open the Timelines tab.

2. Select the desired timeline.

13

3. Choose Window | Properties to open the Properties palette, shown next:

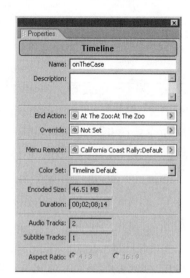

4. Click the triangle to the right of the Color Set field and choose the desired color set from the drop-down list. After you select a color set, Adobe Encore DVD displays a Progress window displaying the text "Apply Subtitle Changes." A progress bar moves from left to right indicating the status of the operation. If you're changing the color set for a long timeline with many subtitles, this process will take a while.

Set Subtitle Clip Position and Display Duration

You can move individual subtitle clips by dragging them to different positions on the timeline. Note that in order to move a subtitle clip, there must be blank space adjacent to the clip in the direction you want to move the clip. You can also manually set the display duration for the subtitle clip. To manually set the subtitle clip duration, follow these steps:

1. Open the Timelines tab.

2. Double-click the desired timeline to display it in the Timeline window.

3. Select the Selection or Direct Select tool.

4. Select the subtitle clip whose duration you want to change.

5. Move your cursor towards the head or tail of the clip.

6. When your cursor becomes a red end bracket with a pointing arrow, drag the clip to increase or decrease the duration of the clip.

 If you are increasing the duration of a clip, there must be empty frames adjacent to the clip.

Edit Subtitle Clip Duration in the Monitor Window

You can also edit the duration for an individual subtitle clip with frame-by-frame accuracy by trimming the clip in the Monitor window. When you trim a clip in the Monitor window, you set the in and out points for the clip as follows:

1. Open the Timelines tab.

2. Double-click the timeline that contains the subtitle clips you want to edit. Adobe Encore DVD displays the timeline in the Timelines window and the first frame of the timeline in the Monitor window.

3. Move the Current Time Indicator over the subtitle clip you want to trim. The subtitle clip is displayed in the Monitor window, shown next:

Trim Subtitle In Point to Here

Trim Subtitle Out Point to Here

13

4. Use the Monitor window controls or the Current Time Indicator to move to the frame where you want the subtitle text to appear.

5. Click the Trim Subtitle In Point to Here button.

6. Use the Monitor window controls or the Current Time Indicator to move to the frame where you want the subtitle text to disappear.

7. Click the Trim Subtitle Out Point to Here button.

Set Subtitle Clip Properties

As mentioned previously, each timeline color set has three groups. You can assign a different color group to each subtitle clip on a timeline. You can also specify the stroke (outline) for a subtitle clip's text, as well as its alignment by modifying a subtitle clip's properties. To modify subtitle clip properties, follow these steps:

1. Open the Timelines tab.

2. Double-click the desired timeline to display it in the Timeline window.

3. Select the subtitle clip whose properties you want to change.

4. Choose Window | Properties to open the Properties palette. When you open the Properties palette for a selected subtitle, the palette is configured, as shown next:

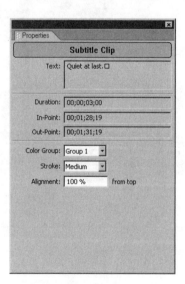

5. Click the triangle to the right of the Color Group field and choose the desired group from the drop-down list.

6. Click the triangle to the right of the Stroke field and choose the desired option from the drop-down list. Your choices are None, Light, Medium, Bold, and Heavy. This property determines the thickness of the subtitle text stroke (outline). Choosing a heavier stroke will make the text stand out from the background.

7. Enter a value for alignment. This option determines where the subtitle text appears in reference to the top (horizontally oriented text) or the right (vertically oriented text) border of the subtitle text box.

After changing the properties of a subtitle text clip, Adobe Encore DVD updates the clip in the Monitor window.

Edit Subtitle Text Characteristics

When you import a subtitle text script or create your own subtitle text in Adobe Encore DVD, you can edit the text characteristics of each subtitle clip to suit your project. For example, you can change the font size to emphasize a subtitle clip. You can also modify other characteristics such as leading, or kerning. For more information on leading and kerning, refer to the information on the Character Palette presented in Chapter 8. To edit subtitle text characteristics, follow these steps:

1. Open the Timelines tab.

2. Double-click the timeline whose subtitle text characteristics you want to modify. Adobe Encore DVD displays the timeline in the Timeline window and opens the Monitor window.

3. Use the Selection or Direct Select tool to select the desired subtitle clip.

4. Choose Window I Character to display the Character palette.

5. Use the Text tool to select the text whose characteristics you want to change.

6. In the Character palette, choose the desired options for the selected text. Note that if you significantly alter the text size, you may have to resize the subtitle text box.

Manage Subtitle Tracks

When you have a production with multiple subtitle tracks, you should specify the first subtitle track as the default or choose Off if you want to give viewers the option of whether or not to display subtitles. When you set up your subtitle tracks, make sure that the tracks on each timeline correspond to each other. For example, if your DVD project has French subtitles, make sure they reside on the same subtitle track in each timeline.

Create a Setup Menu

When you have a DVD project that uses only alternate audio tracks, you set up a menu that DVD viewers use to select audio tracks, as outlined in Chapter 12. When you create a DVD that has multiple subtitle tracks, you can create a similar menu. However, when your project features both multiple audio tracks and multiple subtitle tracks, you should consider creating a setup menu. You've probably seen setup menus in DVD productions of Hollywood movies. The setup menu enables viewers to select both audio and subtitle options. When you set up a menu to select only audio or subtitle selections, you link back to the main menu after the selection is made. However, when you have a setup menu for both options, each button links back to the default button of the setup menu. At the bottom of a typical setup menu is a link to the main menu, as shown in Figure 13-4.

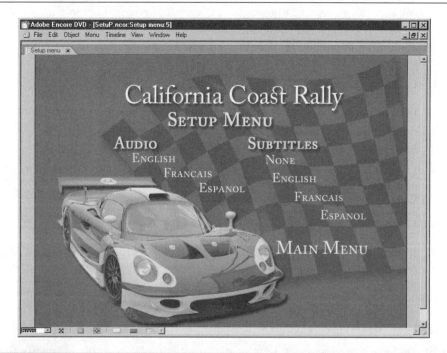

FIGURE 13-4   A setup menu enables viewers to choose audio and subtitle options.

How to ... Create Subtitles in a Word Processing Application

Creating subtitles in Adobe Encore DVD is a fairly straightforward process. However, if you are creating subtitles in multiple languages, you may find it easier to create your own subtitle scripts and them import them in to Adobe Encore DVD. You can create a subtitle script in your favorite word processing application as long as the application has the option to save the file as a Text (*.txt) file. To create a subtitle script, follow these steps:

Open the desired timeline in a video editing application that enables you to preview the timeline in a monitor window. Make sure that the application has a monitor window that displays the current time in hours, minutes, seconds, and frames. Adobe Premiere Pro's Monitor window has this feature. For that matter, so does the Monitor window in Adobe Encore DVD.

1. Play the video and pause playback when you encounter a point at which you'd like to display a subtitle.

2. Note the timecode in the following format hh:mm:ss:ff, where h is hours, m is minutes, s is seconds, and f is frames. For example, 00:00:10:28 is 10 seconds and 28 frames into the video.

3. Continue in this manner until you've noted the timecode for every place in the video where you want a subtitle to appear.

4. Create a blank document in your favorite word processing application.

5. Type the number 1, followed by a space and then the timecode where you want the first subtitle to appear.

6. Type a space, and then the timecode where you want the first subtitle to disappear. For example, if you want the subtitle to disappear in three seconds, add three to the timecode in Step 5.

7. Type a space and then the subtitle.

8. Press ENTER.

9. Type the number 2 followed by the beginning and ending timecode, and then followed by the subtitle text. Make sure to leave spaces as outlined previously.

10. Continue in this manner until you've written every subtitle.

13

11. Save the document as a Text (*.txt) file. The following illustration shows a typical subtitle text being created in Microsoft Word:

The beauty of creating subtitle scripts in a word processing application is that you can spell check the document, a handy feature when you've got a long video with lots of subtitles.

An added bonus is when you have subtitles in multiple languages as each subtitle will appear at the same place in the production. Open your first subtitle document; replace the text for each subtitle with the desired language, and then save the document as a text file with a different name. You can then import each subtitle script and add it to its appropriate timeline subtitle track. Note that when you create a subtitle script for multiple languages, you should create the subtitle text in a text editor such as Notepad and save the document as a UTF-8 text file.

Create Setup Menu Choices

When you create a setup menu, you use a menu with the same background image, text fonts, and other styling attributes as the menus in the rest of your project. The easiest way to begin creating a setup menu is to duplicate the main menu in your project. Open the duplicate in the Menu Editor and then revise the button text to describe each audio and subtitle choice available. To keep things neat and tidy, give the setup menu a unique name (*Setup Menu* is a good choice) and rename each button as well.

You can rename a button by selecting it, opening the Properties palette, and then enter a new name in the Name field. You may have to duplicate buttons in order to flesh out your setup menu. After all the buttons are created, you can create the button links.

Link to Subtitle Tracks

After you create the buttons for your setup menu, the next step is to set up the links for each audio button, as outlined in Chapter 12. Then you'll create the links for your subtitle buttons as follows:

1. Open the Menus tab.

2. Double-click the setup menu to open it in the Menu editor.

3. Select the desired button.

4. Choose Window | Properties to open the Properties palette.

5. Click the triangle to the right of the Link field and choose Specify Other to open the Specify Link dialog box.

6. Create a link to setup menu's default button.

7. Click the triangle to the right of the Subtitle field and choose an option from the drop-down list, as shown next:

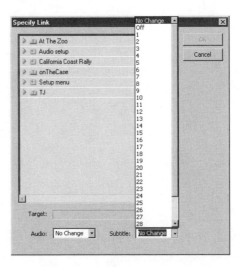

8. Repeat Steps 3 through 7 for the rest of the subtitle buttons.

After you create a link for every subtitle button in your setup menu, it's a simple matter to create a link for the main menu button, which will display the DVD project's main menu after viewers choose the desired audio and video tracks.

Summary

In this chapter, you learned to add subtitles to a DVD project. Specifically, you learned how to add subtitle tracks to video timelines and how to import subtitle images and scripts, as well as create your own subtitles in Adobe Encore DVD. You learned to create timeline color sets for subtitle text display, edit subtitle properties, and trim the duration of subtitle clips. In Chapter 14 you'll learn how to prepare your project for final output.

Part V

Create the DVD

Chapter 14

Prepare the Project for Output

How To...

■ Use the Project Preview window

■ Test links

■ Copy protect a disc

■ Set region coding

After you import assets, create timelines, set chapter points, and create menu navigation, you may think you're ready to build the project. However, your best efforts are for naught if any part of your project doesn't perform as envisioned. You can save yourself a lot of grief (not to mention useless DVD discs) by making sure that all of your links function perfectly, and that there are no orphaned menus or timelines. When you're preparing the project for final output, you can also apply any copy protection to the disc.

In this chapter, you'll learn the final steps necessary to prepare the project for final output. You'll learn to preview the entire project, as well as track down broken links and orphaned timelines. You'll also learn how to apply copy protection to discs and how to set region coding. At the end of the chapter, you'll find a handy prebuild checklist.

Preview the Project

Before you actually build a project, you should preview the project to make sure all menu buttons function properly, that timeline end actions execute properly, that the menu and title button links are correct, and so on. You preview a project in the Project Preview window. You received an introduction to the project preview window in Chapter 9. In the sections that follow, you'll learn to use all the features of the Project Preview window. To preview your entire DVD project, choose File | Preview. Adobe Encore DVD launches the project in the Project Preview window (shown in Figure 14-1) beginning with the project's First Play.

 *If you're previewing a project with motion menus and want to preview the menus in action, choose File | Render Motion Menus to build the motion menus before previewing the project.*

Navigate the Project Preview Window

When you preview a project, you use the controls in the Project Preview window just as you would a remote control device for a set top DVD player. In addition

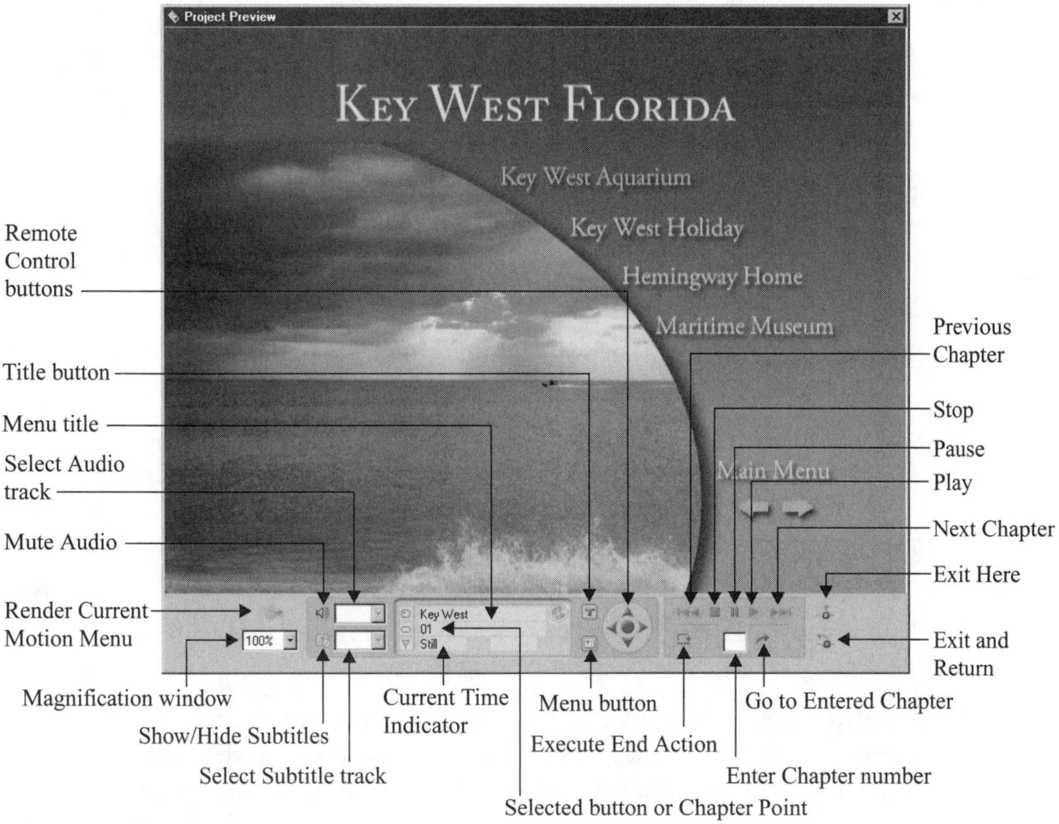

Remote Control buttons

Title button

Menu title

Select Audio track

Mute Audio

Render Current Motion Menu

Magnification window

Show/Hide Subtitles

Select Subtitle track

Current Time Indicator

Menu button

Execute End Action

Selected button or Chapter Point

Previous Chapter

Stop

Pause

Play

Next Chapter

Exit Here

Exit and Return

Go to Entered Chapter

Enter Chapter number

FIGURE 14-1 You use the controls in the Project Preview window to test buttons, end actions, and so on.

to the remote control buttons, you have other buttons available for testing end actions, playing specific chapters, and so on. For example, you can navigate to specific chapter points, mute audio, select an alternate audio track, and so on. The following list explains the controls in the Project Preview window, previously shown in Figure 14-1.

- **Render Current Motion Menu** Renders the current motion menu.

- **Magnification window** Displays the current magnification of the menu. Click the triangle to the right of the field, and you can reduce the magnification to 50 percent by choosing that option from the drop-down menu. You cannot manually enter a value in this field. Your magnification options are 50 or 100 percent.

- **Mute Audio** Toggles audio on and off.

When you click the Mute Audio button, Adobe Encore DVD takes control of your system sound and mutes the volume. If you click the Mute Audio button when previewing a menu, be sure to click it again after you finish previewing the menu; otherwise, your system sound will be disabled when you exit Adobe Encore DVD.

- **Select Audio track**   Displays the number of the audio track currently playing. If you're playing a timeline with multiple audio tracks, you can listen to a different track by clicking the triangle to the right of the field and choosing the desired track from the drop-down list.

- **Show/Hide Subtitles**   Displays or hides the text subtitle clips on the subtitle track(s).

- **Select Subtitle track**   Displays the number of the subtitle track currently being displayed. If you're playing a timeline with multiple subtitle tracks, you can display the subtitle text on a different subtitle track by clicking the triangle to the right of the field and choosing the desired track from the drop-down list.

- **Title window**   Displays the title of the currently selected menu, title of the timeline, or chapter point currently being played.

- **Selected button or Chapter Point**   Displays the number of the currently selected button or the chapter currently playing. The icon to the left of the field changes to reflect whether the displayed text is for a button or chapter point.

- **Current Time Indicator**   Displays the current time of a motion menu or the current time of the timeline/chapter point being played. If a static menu is displayed, the word *Still* appears in this window.

- **Audio type**   Displays the audio codec for the audio track being played with a timeline.

- **Remote Control Title button**   Displays the project title menu.

- **Remote Control Menu button**   Displays the menu linked to the timeline/ chapter point end action.

- **Remote Control buttons**   The arrow buttons are used to navigate to menu buttons. The round button in the center functions like the Enter button on a set top DVD remote controller.

- **Previous Chapter**   Plays the previous chapter when clicked.

- **Stop**   Stops the current timeline from playing.

■ **Pause**   Pauses the current timeline when clicked; resumes play when clicked again.

■ **Next Chapter**   Plays the next chapter when clicked.

■ **Execute End Action**   Executes the timeline or menu end action when clicked. Be careful not to use this button right after a timeline begins playing as the application may stop responding. Allow the timeline to play for several seconds before using this button. This gives the application and your system time to stabilize.

■ **Enter Chapter Number**   Enter a chapter number that you want to preview.

■ **Go to Entered Chapter**   Plays the chapter specified in the Enter Chapter Number field when clicked.

■ **Exit Here**   Stops the timeline or current menu and exits the project preview. Adobe Encore DVD opens the applicable window for the object that was displayed in the Project Preview window when you clicked the Exit Here button. For example, if you're previewing a menu, when you click the Exit Here button, Adobe Encore DVD opens the menu in the Menu Editor, giving you the opportunity to edit the menu.

■ **Exit and Return**   Stops the current timeline or menu, exits the project preview, and returns you to the window or tab you were working in when you began previewing the project.

When you're previewing your project, you should check each button to make sure it links to the proper timeline. If your project contains motion menus, make sure they loop properly. Remember, if you forget to set looping on a motion menu, the end action for the menu executes immediately after the background video ceases playing. If the end action is the default, Stop, your viewers will be confronted with a black screen. You should also be sure that the Menu and Title buttons link to the desired menus.

If you don't have time to watch an entire timeline, you can always click the End Action button to ensure the proper menu or timeline appears. If you click a button to which an override action has been assigned, after the associated timeline begins playing, you can click the End Action button to test the override—the override action executes in place of the timeline end action. However, if you've assigned unique names to each menu and timeline, you can also check the Menus tab and check the Override column for each button. If the proper menu link appears in that column, you're good to go.

Building a DVD disc is a lengthy process. Spending a few minutes previewing your project is cheap insurance, especially if you're going to have a replication service

14

create several hundred, or several thousand, copies of your project. About the only thing a failed DVD disc is good for is a beverage coaster, albeit a very expensive beverage coaster. Of course, as mentioned previously, you can always build the project to a rewritable disc. If you do so and happen to notice anything amiss while previewing the project, you can always right your wrong and once again build the corrected project to the rewritable disc.

Test for Missing Links or Unused Assets

After you preview your project, you should also test for any missing links you may have overlooked during the preview process. You can also check for unused assets, such as unused menus or timelines. Remember that all menus and timelines are added to the project build, regardless of whether or not you've used them.

Adobe Encore DVD automatically checks for broken links when you build a project. You can rely on Adobe Encore DVD to find any broken links before building the project. However, the best insurance you have is to test links prior to building the project. When you test links, you may find that some buttons link to a timeline you've replaced. You may also find you've deleted one or more buttons from your project and as a result have timelines that cannot be viewed. Of course, if your viewers have DVD players with a title search feature, they can search the disc for all titles. However, a DVD disc with timelines that cannot be accessed by menus cannot be considered the work of a professional.

To check the links and assets in your project, follow these steps:

1. Choose Edit | Check Links. Alternatively, you can click the Disc tab, and then click the Check Links button. Either method opens the Check Links dialog box, shown next:

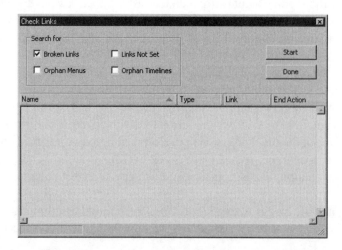

2. Select one or more of the following search options:

- ■ **Broken Links**   Checks your project for links that point to missing buttons, menus, or timelines.

- ■ **Links Not Set**   Checks your project for links that are not set, such as end or override actions.

- ■ **Orphaned Menus**   Checks your project for menus that have been added to the project but have not been used.

- ■ **Orphaned Timelines**   Checks your project for timelines to which links have not been set.

3. Click Start. Adobe Encore DVD searches for the selected options and returns the results in the lower half of the Check Links dialog box, as shown next. Note that the Description column has been hidden in this illustration so that you may view the other columns.

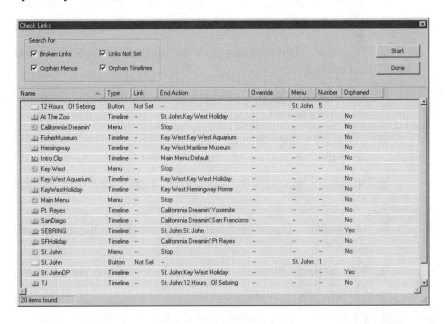

The results columns in the Check Links dialog box function in a similar manner to those in the Project, Timelines, and Menus tabs. Notice that there are several different types of assets listed in the Type column as all options were selected. All timelines are displayed to alert you whether or not the timeline is orphaned when the Check Links button was clicked. Also notice that there are two buttons for which links have not been set and two orphaned timelines. When you see similar results when checking a project for broken links and orphaned timeline, it's a dead giveaway that the buttons should link to the timelines.

You can choose which columns to display by right-clicking any column title to reveal the shortcut menu and then selecting the column you want to display or hide. Displayed columns are designated by a checkmark to the left of the column name. You can hide any column by right-clicking the column name and then choosing Hide This from the shortcut menu.

You can sort the results of a column by clicking the desired column's name. An up-pointing arrow in the column signifies that the results are sorts in ascending order, while a down-pointing arrow signifies the results are sorted in descending order. Clicking the column title toggles the sort order.

The columns in the Check Links dialog box are as follows:

- **Name**   Displays the name of the suspect asset.

- **Type**   Displays the type of asset: menu, button, or timeline.

- **Link**   Displays Not Set if the link for a button has not been set.

- **Override**   Displays Not Set if the override action for a timeline has not been set.

- **Description**   Displays the description for the item as entered in the Properties palette.

- **Menu**   Displays the menu on which a button with an unset link can be found.

- **Number**   Displays the number of a button for which a link has not been set.

- **Orphaned**   Displays Yes if the timeline or menu is orphaned, No if it is not.

- **End Action**   Displays Not Set if the end action for the timeline or menu has not been set.

4. To correct a broken link, unset link, unset end, or override action, select the item in the Check Links dialog box; next, open the Properties palette and set the desired link or action.

5. To create a link to an orphaned menu or timeline, determine which item should link to the orphaned item, open the Properties palette, and set the link.

6. Repeat Steps 4 and 5 for other broken links or orphaned assets.

7. Click Done to exit the Check Links dialog box.

If your search returns any orphaned timelines or menus that you don't want to include in the project, open the Project tab and delete the unwanted item. If you leave it in the project, it will be included in the build and use up valuable disc space. If the orphaned item is a timeline, you run the risk of gobbling up a huge amount of disc space. If your assets have already been transcoded, there may not be enough room for the project until you delete the unwanted orphan. If the assets have not been transcoded, Adobe Encore DVD will have to use higher compression in order to fit everything on the disc, which may translate to poor video quality.

Finalize Project Settings

Prior to building your project, you can finalize settings for region coding and copy protection. By default, copy protection is disabled unless you change Encoding Preferences as outlined in Chapter 3. Copy protection that is set with Encoding Preferences is applied to all subsequent projects. You can also set copy protection for a project on which you are currently working. The upcoming sections show you how to set region coding and copy protection for your projects. You'll also gain an understanding of how copy protection works. To set region coding and copy protection, you work in the Disc tab. You open the Disc tab, shown next, by clicking its tab or by choosing Window | Disc:

14

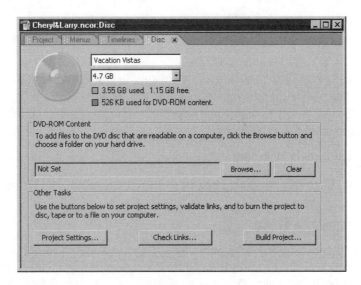

Set Region Coding

By default, a DVD project you create with Adobe Encore DVD can be played on set top DVD players from all regions. You can however, limit the playback of your project to DVD players sold in specific regions. To set region coding for a DVD project:

1. Open the Disc tab, as outlined previously.

2. Click the Project Settings button to reveal the Project Settings dialog box, shown next.

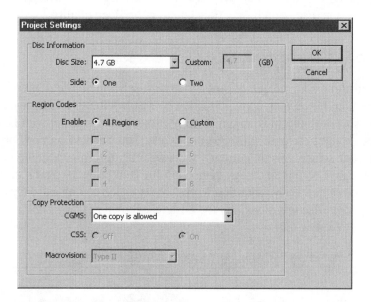

3. In the Region Codes section, click Custom to enable the region code radio buttons. Notice that the region in which you reside is selected by default and grayed out. The only way to deselect your own region code is by first selecting another region code.

4. Select the desired regions.

5. Click OK to exit the dialog box or to set a different project setting.

Copy Protect a Disc

If you've set preferences for copy protecting projects using the Master Preferences dialog box, they are automatically applied to subsequent projects. Unless all of your projects are going to be built to DLT tapes, your best bet is to leave Master Preferences

Did you know?

About Copy Protection

When you copy protect a disc, you limit or prohibit copying of your project. When you copy protect a project, the project must be built to a DLT (digital linear tape), which is then used to replicate the disc. The copy protection you specify is applied to each replicated disc.

When you copy protect a disc, you can allow users to create one copy or none. If you do not copy protect project, users can make unlimited copies of the DVD disc. Adobe Encore DVD gives you three tools to use when copy protecting a disc. They are as follows:

- **CGMS (Copy Generation Management System)**   This option allows you to restrict the number of copies that can be made of the disc.

- **CSS (Content Scrambling System)**   This option encrypts the video data. Keys to decrypt the video are included with the original disc. These keys, however, cannot be copied to another disc.

- **Macrovision**   This option also encrypts the video data. Furthermore, the video cannot be copied to an analog device such as a VCR player. Macrovision sends an electronic pulse with the video data that effectively scrambles the data when recorded by an analog device. There are three types of Macrovision, all of which are available to you with Adobe Encore DVD:

  - Type I Macrovision affects the AGC (automatic gain control) of the analog device attempting to record the DVD. The resulting video has widely varying degrees of brightness, making it unviewable.

  - Type II Macrovision affects the AGC of the analog device and adds two lines of disturbance to the video signal.

  - Type III Macrovision affects the AGC of the analog device and adds four lines of disturbance to the video signal.

However, when you use Macrovision to copy protect a project, there is a per disc royalty you will be required to pay when you have the project replicated. The replication facility can give you more information about the royalty.

14

at their default settings and add copy protection when building a project to DLT tape. If you change Master Preferences from copy protection to no copy protection, unless you deselect each and every option in the Master Preferences dialog box, you may have copy protection enabled without knowing it. Therefore, you should set copy protection for each project that needs it instead of relying on Master Preferences; otherwise, the projects you burn to DVD discs may have errors. To copy protect a project, follow these steps:

1. Open the Disc tab as outlined previously.

2. Click the triangle to the right of the CGMS field and choose one of the following options from the drop-down list:

   ■ **No Copies Are Allowed**   This option prohibits copying of the disc.

   ■ **One Copy Is Allowed**   This option allows one copy of the disc to be made.

   ■ **Unlimited Copies Are Allowed**   The default option allows unlimited copies of the disc to be made.

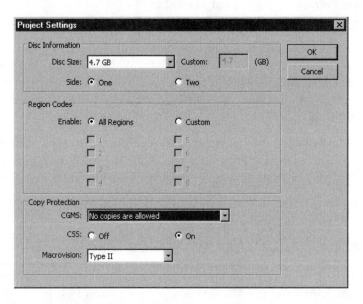

3. For the CSS option, choose Off or On.

4. Accept the default Macrovision option of Off or click the triangle to the right of the field and choose Type I, Type II, or Type III.

5. Click OK to exit the dialog box or set a different project setting.

Archive Project Files

After you've previewed a project and prepared everything for output, you may want to consider archiving the project files for future use. If you're working for a client who may want you to revise the project in the future, working from an archived version of the project is certainly easier than starting from scratch. You can archive the files to DVD disc as outlined in Chapter 5, or you can save the project and all associated files to a mass storage device such as an external USB or Firewire hard drive. Another option is saving the files to a tape backup. When you archive the project, copy the entire project folder, the associated audio, the image and video files, and the project NCOR file to the desired device. Make sure you keep the same relative path to the audio, image, and video assets on your archive device as you have on your system hard drive. When you need to revise the project, simply copy the project folder, assets, and NCOR file to your hard drive. Launch Adobe Encore DVD and open the NCOR file—you have access to all the files you need to revise the project.

Use a Prebuild Checklist

There's nothing more frustrating than committing a DVD project to disc only to find out that the DVD project doesn't play as desired when you test it on a set top DVD player or on a computer with DVD software. Of course, you can always build a prototype of the project to reusable DVD media. Even so, building a 4-gigabyte DVD takes a long time and uses a considerable amount of system resources. To make sure you get it right the first time, you may want to consider creating a prebuild checklist similar to the following:

- Has each facet of the project been previewed in the Adobe Encore DVD Project Preview window?

- Does every menu button link to the proper source?

- Do the menu end actions function as desired?

- Does each timeline end action link to the proper source?

- If your menu buttons use override actions, do they link to the proper menu?

- Will the project fit on the desired media? This issue is a concern if you've imported DVD-legal MPEG-2 files that have been rendered in another application. If you've imported AVI files, Adobe Encore DVD will automatically determine the proper transcoding to fit the project on the specified media.

14

■ If your project has motion menus, do they loop as desired?

■ If your project has static menus with animated buttons, do the buttons loop for the desired duration? Do they loop for the desired number of repetitions?

■ If your project uses motion menus, animated buttons, and background audio tracks, are they synchronized as desired? Remember, these assets play for the specified menu duration. If the audio or video background tracks are longer than the menu duration, they will not play completely, which may produce undesirable results.

■ Have you tested for broken links or unused assets? Adobe Encore DVD will alert you to any broken links prior to building the project, but it's always best to tend to these issues before you build the project, especially if you're under a tight deadline.

■ Are the subtitles synchronized with the video? (For more information on subtitles, see Chapter 13.)

■ Have you provided menus to play alternate audio and subtitle tracks? If so, does the proper audio or subtitle track play when the button is selected?

Summary

In this chapter, you learned to prepare your project for output. You learned to preview your project in the Project Preview window to ensure that all objects in your project function as desired. You also learned to test your project for broken links and unused assets as well as set copy protection and select region coding. In the next chapter, you'll learn how to build your project.

Chapter 15

Build the DVD

How To...

■ Select media

■ Choose your build settings

■ Create DVD folders and images

■ Build a DVD disc

After you preview your project and test for broken links, unused timelines, and orphaned menus, you're ready to build the project. When you build a project, you have options. You can build a DVD disc, create a DVD image file create a DVD master, or create a DVD folder. Each option has its place. Image files and DVD folders are generally used when you're going to replicate several copies of a disc in-house. You create a DVD master when you add copy protection to a project, which you then outsource to a company that replicates the desired number of discs.

In this chapter you'll learn how to build a project to a DVD disc. Your first step will be to review your build settings. After you ensure the build settings are in order, you'll set the build options. After tending to those DVD housekeeping tasks, you can tend to other tasks such as making sure your file cabinet is in order or your desk is clean, because your computer will be tied up for several minutes—or perhaps an hour or more depending on the size of the project you're building, the type of media to which you're burning the project, and the speed of your DVD burner.

Review Project Settings

As a rule, you specify project settings after you create a new project. However, sometimes things change. If the client presents you with additional material or decides to offer a widescreen version of a movie, a project that began as a single-sided disc becomes a dual-sided disc. Even if nothing has changed since you created the project, it's a good idea to check your settings prior to committing the project to disc. To review project settings, follow these steps:

1. Open the Disc tab.

2. Click the Project Settings button to open the Project Settings dialog box, shown next:

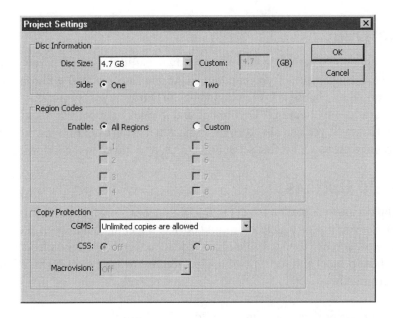

3. Review the settings and make any necessary changes.

At this time it's also a good idea to review region settings and copy protection settings. If you're building a project to DVD disc, it's a good idea to disable copy protection settings. Copy protection is designed for a project that will be written to DLT tape, which will be used to replicate the disc. Although a DVD disc with copy protection may function perfectly, the copy protection will not work; therefore, it's best to leave copy protection with the default option of unlimited copies.

CAUTION   *If you change Mastering Preferences (Edit | Preferences | Mastering) with the intention of disabling copy protection, you must change the CGMS option to Copy Permitted Without Restriction (Off), change the Macrovision option to Off, and click the CSS Off radio button before deselecting the Copy Protection option. If you fail to disable CGMS or Macrovision before deselecting the Copy Protection option, CGMS or Macrovision copy protection will still be applied to future projects you create. In other words, when you deselect Copy Protection, it does not set the CGMS and Macrovision settings to their default states.*

15

Build the Disc

After reviewing project settings, you're ready to build the project. You can build using the current project, or you can launch Adobe Encore DVD without opening a project and then choose a DVD volume or DVD image you've previously built. (You'll learn how to create DVD volumes and images in upcoming sections of this chapter.) However, you can also burn the project to a DVD disc. Before you can burn the project to disc, you must first set build options, as outlined in the following section.

Set Build Options

When you decide to commit a project to DVD disc, you must first make a few choices. For example, you'll need to select the build device, select the write speed, and specify the number of copies. The available choices differ depending on the media to which you build the project and the type of device you use to burn the DVD. To build a DVD, follow these steps:

1. Open the project you want to build to DVD disc.

2. Choose File | Build DVD | Make DVD Disc. Adobe Encore DVD displays the Save Project dialog box (shown next) prompting you to save the file prior to building it.

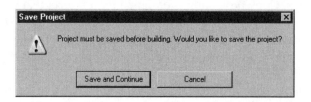

3. Click Save and Continue to open the Make DVD Disc dialog box, shown next. Note that by default, the Source section already indicates the current project. Adobe Encore DVD has also selected your system's default device for building DVD discs.

NOTE *You can also build a disc from a DVD volume (folder) or a DVD image by clicking the triangle to the right of the Create Using field and then choosing the desired choice from the drop-down list. Select the desired DVD volume or image file before proceeding to the next step.*

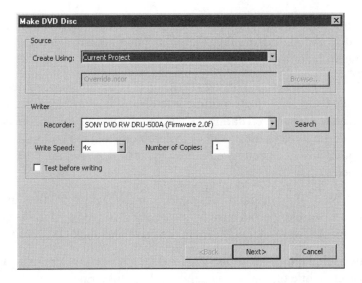

4. Accept the default build device, or if you have more than one device capable of burning DVD discs, click the triangle to the right of the field and select the desired device from the drop-down list. Alternatively, you can click the Search button and Adobe Encore DVD will search for additional devices.

5. Accept the default Write Speed or click the triangle to the right of the field and choose the desired speed from the drop-down list, as shown next. Note that the speed selected is the maximum speed at which your device can write a DVD project to the media currently in the DVD drive.

15

6. Enter a value in the Number of Copies field. If you burn multiple copies of the disc, Adobe Encore DVD prompts you to insert a new disc when the previous build is finished.

7. Click the Test Before Writing checkbox if you want Adobe Encore DVD to test the build prior to committing it to disc. If any problems such as faulty media are encountered during the test, Adobe Encore DVD displays a dialog box alerting you to the nature of the problem, and the build is cancelled. If no problems are encountered during the testing phase, the project will be burned to disc.

8. Click Next. If no media is present in your DVD drive, Adobe Encore DVD opens the disc drive and displays a warning dialog box. If this occurs, place a blank or rewritable DVD disc in the drive and click OK.

9. After doing either of the previous, Adobe Encore DVD begins building the project. While the project is being built, Adobe Encore DVD displays the following dialog box, which informs you which stage of the build is occurring:

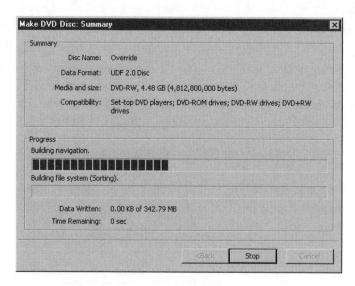

It takes a considerable amount of time to burn a DVD disc, especially if you're building a project with a lot of video. The build time also depends on the type of media to which you are building the project. The longest build time occurs when you build the project to DVD-RW discs, while the quickest occurs when writing a project to DVD+R discs.

Adobe Encore DVD ejects the disc immediately after the lead out has been written. The following dialog box is displayed, informing you that the project has been successfully been built to disc. Click OK to exit the dialog box and resume working:

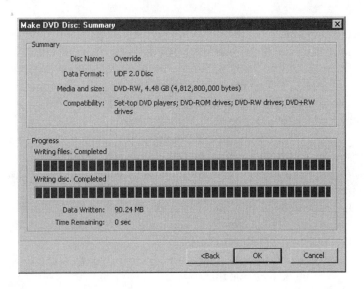

CAUTION

While the DVD disc is being burned, you will not be able to use Adobe Encore DVD for any other purpose. In fact, the process of burning a DVD is processor intensive to the extent that you actually run the risk of crashing your computer if you try to multitask. Therefore, you should not use your computer for anything else while the disc is being created. You may also find it necessary to reboot your system after the disc has been built.

NOTE

You can also build a project by clicking the Build Project button in the Disc tab.

15

Building Discs with MPEG-2 Sound

Many set top DVD players require that at least one AC-3 or PCM audio track be available per NTSC title track. If any of your titles do not have an AC-3, Adobe Encore DVD displays the following warning dialog box when you build the project:

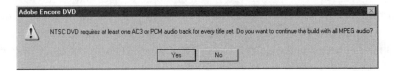

Mass Produce DVD Discs

If you create DVD discs for distribution to clients or customers, you need some method of creating copies of the discs. If you're only creating a few copies at a time, you can use Adobe Encore DVD or 3rd party software to duplicate the disc. If you have to create a large quantity, you can hire a firm to duplicate the discs for you. The firm can also create labels for the DVD discs from your artwork or prepare the artwork to your specifications in-house.

If you frequently update your DVD discs to reflect changes in your product offering or services offered, it may be economically feasible to invest in a DVD duplicator. DVD duplicators range in price from several hundred dollars for a one-to-one duplicator to several thousand dollars for machines that can create several duplicates at the same time. Many DVD duplicators contain hard drives that you use to store DVD Images. As an added bonus, you can also use the machines to duplicate CD-ROM discs.

You can also invest in hardware to print labels for your DVD/CD-ROM discs. There are also printers available that you can use to print labels directly to DVD or CD-ROM discs. These printers require you to use discs with special faces to which images can be printed.

You can find many sources for DVD and CD-ROM duplicators on the Internet. Go to your favorite search engine and type **DVD Duplication Hardware** in the search field. You'll find a wide variety of sources where you can obtain prices and other information.

While the hardware is expensive for a casual user, if your firm frequently needs to duplicate small quantities of new material, this may be the ticket. If your requirements are less than a duplication firm's minimum order, but your frequency of duplication is high enough, the machines will pay for themselves in convenience alone. It's important to note that it is illegal to duplicate any DVD or CD-ROM disc that contains material for which you do not own the copyright.

Even if you build the disc with MPEG-2 sound, it will work fine on most newer NTSC set top DVD players. However, if you build the project and intend to distribute the resulting disc to a large audience, the sound may not be compatible with older NTSC DVD players. Therefore, when you see this warning, you may want to consider transcoding the file using the default Dolby Digital sound encoding or PCM encoding.

Write DVD-Compliant Files to CD-ROM Discs

If you have a small project that totals less than 700MB, you can build the project to CD-ROM discs for distribution to a viewing audience that has software to play DVDs on their PCs. When you build a project to CD-ROM, you can also include material that can be viewed on a computer, such as PDF files. You will not be able to include button links on your DVD menu to access computer readable material; therefore, you'll have to provide instructions to your viewing audience as to which folder the material can be found in, what software they'll need to view the files, and so on. When you intend to include computer-readable files in a DVD project, make sure you include the desired files in their own folder as you can add a folder only to the project to be built and cannot add individual files to the project. To build a DVD project to a CD-ROM disc, follow these steps:

1. Open the project you want to build to a CD-ROM disc.

2. Open the Disc tab shown here:

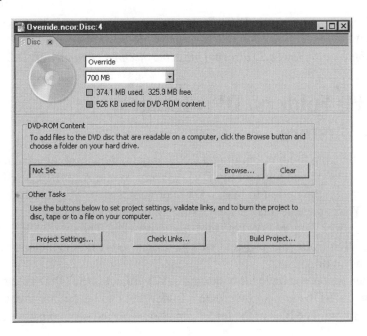

3. Make sure you've specified the proper disc size. You should have either 650 or 700MB specified for the media size. At the same time, check to make sure that the project will indeed fit on the specified disc size. This information is displayed below the specified media size.

4. In the DVD-ROM content section, click the Browse button to open the Browse for Folder dialog box.

5. Select the desired folder and click OK to add it to the project. The path to the folder appears in the DVD-ROM content section, and the disc space used and remaining is updated. The DVD-ROM listing also increases to reflect the folder's overhead on the project.

6. Click Build Project. Adobe Encore DVD displays the Save Project dialog box.

7. Click Save and Continue. Adobe Encore DVD displays the Build DVD disc. Follow the steps outlined previously in the "Build the Disc" section of this chapter to complete the build.

 *You can also include DVD-ROM content on a project that you will build to DVD disc. The content will be accessible by anyone whose computer has software to play DVD disc and plays the disc in a computer DVD-ROM drive.*

Create DVD Folders, DVD Images, and DVD Masters

When most people think of building a DVD project, they think of creating a DVD disc that will be played with set top DVD players. While Adobe Encore DVD does indeed give you the option to burn a DVD project to disc, you have other options available. You can create a DVD image file, which can then be used to create a DVD disc using Adobe Encore DVD or third-party software. The image file contains all the information from your project that's necessary to create the menus, timelines, buttons, and so on.

Another option you have for building a DVD project is the DVD volume. When you create a DVD volume, you create a folder that can be used to create a DVD disc in the future. The files in the folder can also be played on a computer equipped with software to play DVD files. If your system is equipped with a DLT drive, you can create a DVD master, which can then be used to replicate the project. In the sections that follow, you'll learn how to create DVD image file, create a DVD master, and create a DVD folder.

Create a DVD Folder

Another build option you have is to create a DVD folder. When you create a DVD folder, you create a folder that can be played on a PC-based software DVD player. When you create the folder, Adobe Encore DVD creates an Audio TS and Video TS folder. The folder also doubles as a DVD Volume from which a DVD disc can be built. To create a DVD Folder, follow these steps:

1. Open the desired project.

2. Choose File | Build DVD | Make DVD Folder. Adobe Encore DVD opens the Save Project dialog box.

3. Click Save and Continue to access the Make DVD Folder dialog box, shown next:

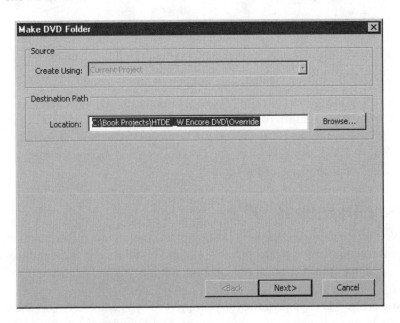

4. Accept the default folder or click the Browse button to specify a different folder.

5. Click Next to display the Summary section of the Make DVD Folder dialog box, as shown here:

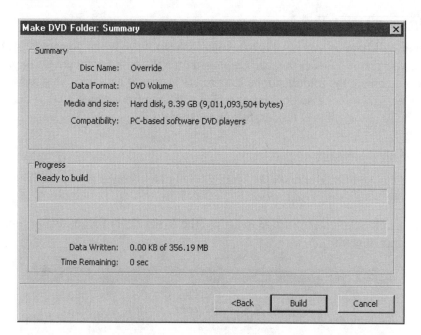

6. Click Build to create the DVD Folder. Alternatively, click Back to display the previous section of the Make DVD Folder dialog box and change the folder in which the DVD Folder is to be created. After you click Build, Adobe Encore DVD begins building the DVD folder, and progress bars indicate the status of each phase of the build.

Create a DVD Image File

When you create an image file of a DVD project, you create a file that contains all the necessary information to create a DVD disc. The advantage of creating an image file is that all the information needed to burn a project to a DVD disc is contained in one file on your system.

If you create a DVD image file for your projects, you can easily sort and organize them on a mass storage device such as an external Firewire or USB hard drive. You can use the image file to burn a DVD to disc as needed. To create a DVD image file, follow these steps:

1. Open the desired project.

2. Choose File | Build DVD | Make DVD Image. Adobe Encore DVD displays the Save Project dialog box, whether or not you've edited the project since opening it.

3. Click Save and Continue. Adobe Encore DVD opens the Make DVD Image dialog box shown here:

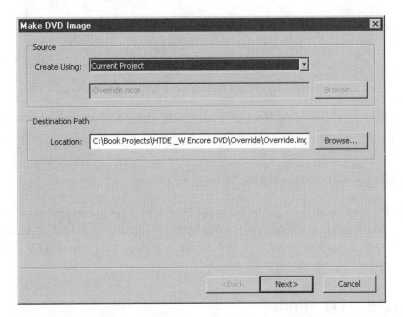

4. Accept the default path and filename for the image file or click the Browse button to open the Save As dialog box, which you use to navigate to the desired folder and enter the desired name for the file.

5. Click Next. Adobe Encore DVD displays the Summary section of the Make DVD Image dialog box, shown next. This dialog box displays information about your project and the drive to which you are saving the image file. If the information is correct, you're ready to build your project as a DVD image. Alternatively, you can click the Back button to navigate to a previous section of the dialog box or cancel the build.

15

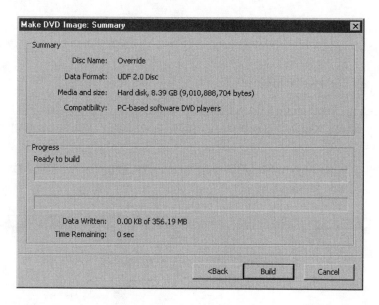

6. Click Build. The dialog box changes to show a progress section. Two progress bars monitor the status of the build. The build goes fairly rapidly, comparatively speaking.

7. Click OK to exit the dialog box.

 If you're building a project that is larger than four gigabytes, you must build the project to a hard drive that has been formatted with the NTFS file system. Hard drives formatted with FAT 32 file system have a maximum file size limit of four gigabytes.

Create a DVD Master

If you've created a project to which you've added copy protection, the only way you'll be able to mass produce discs with copy protection is by creating a DVD master on a DLT tape drive and then outsourcing the replication of the build to DVD discs. To create a DVD master, follow these steps:

1. Open the desired project.

2. Choose File | Build DVD | Make DVD Master. Adobe Encore DVD opens the Save Project dialog box.

3. Click Save and Continue to open the Make DVD Master dialog box.

4. Click Next to build the project to your DLT tape device.

Replicate a DVD

You can use third-party software or Adobe Encore DVD to create multiple copies of your disc. This method works well if you're creating limited copies of a disc. If you need to create several dozen copies of projects on a regular basis, you may want to consider investing in duplication hardware. If you need to create several hundred or several thousand copies of a disc, you can outsource the project to a disc duplication or disc replication facility. For more information on disc replication read the following Did You Know section or the previous "Mass Produce DVD Discs" sidebar.

Did you know?   Understanding Disc Replication

If you choose to outsource disc replication for a DVD project, you have many choices. The least expensive way to go is to have the project duplicated to DVD-R media. When you hire a firm to duplicate a project to DVD-R media, they have the necessary software and hardware to create multiple copies of your project on the same media you use to create one or a few DVD discs using the DVD burner on your system. The hardware used by a duplicating service has the capability to create thousands of discs a day. Most of the hardware is automated—the duplication service loads your master disc into their machine, loads the necessary number of discs on a spindle, and then sets the hardware for the desired number of copies. Most duplication services can also create labels for your disks as well as packaging.

If you need to create several hundred copies of a project to test the waters, so to speak, a duplication service is an economical solution. However, if you're creating thousands of copies of a presentation, you may want to consider a replication service. The difference between a duplication service and replication service is the quality of the final product. A duplication service uses the same type of discs you use to burn DVDs on your computer. A replication service creates a coated glass master of your project, which is then molded into a stamper. The final product is a polycarbonate disc with a reflective surface. A protective coating is applied to the surface, and the label is printed directly to the disc. The end product is more durable and scratch resistant than a duplicated disc. You can readily tell the difference between a duplicated disc and a replicated disc by the playing surface. A duplicated disc has a shiny green or blue surface, whereas a replicated disc has a highly reflective, shiny silvery surface.

Create a Dual-Side or Dual-Layer Disc

When you create a new project for a dual-sided disc, you specify the size of the media and the side of the disc for which the project is created. You cannot build a dual-sided, or for that matter, a dual-layer DVD directly to DVD disc. You must create a separate DLT tape for each layer (in the case of a dual-layer disc) or each side (in the case of a dual-sided disc) of the project. To create the actual discs, you hire a replication facility to convert the DLT tapes into finished discs.

Summary

In this chapter, you learned how to build a DVD project. You learned how to make a DVD disc, a DVD master, a DVD folder, and a DVD image. You also learned about available hardware to mass produce discs along with how to duplicate and replicate discs. In the appendixes that follow, you'll find some useful information pertaining to Adobe Encore DVD and the process of authoring a DVD disc. If you don't use Adobe Encore DVD for an extended period of time, or you need a refresher course on how to author a DVD, Appendix A will do the job nicely. If you like to streamline your workflow with keyboard shortcuts, you'll find all of the Adobe Encore DVD keyboard shortcuts listed in Appendix B. And if you're one of those people who like to augment their knowledge base with a bit of Internet surfing, you'll find a list of web sites that feature DVD and digital video editing information in Appendix C.

Part VI

Appendixes

Appendix A

Create Your Own DVD: A Step-By-Step Tutorial

How To...

■ Capture Video

■ Assemble Assets

■ Plan Your Project

■ Create the Menu

■ Create Menu Navigation

■ Preview Your DVD

■ Build Your DVD

When you author a DVD in a Adobe Encore DVD, you take many steps before you're able to build the finished product. Unlike less sophisticated DVD-authoring applications, Adobe Encore DVD gives you choices. But with these choices come opportunities for errors. For example, if you forget to create an end action for a timeline, your viewers are confronted with a black screen. If you forget to create a link for a button, your viewers are unable to view the video. In order to take advantage of the full capabilities of Adobe Encore DVD, you should adopt a methodical workflow that enables you to create a professional-looking DVD that is flawless.

If you've read through the entire book, you know the importance of planning your project and working in an organized manner. This appendix is designed as a guide that encapsulates all of the steps necessary to create your own DVD. Your path may differ slightly if your project doesn't require subtitles or alternate text. In each of the upcoming sections, there will be references to the chapters where this material is covered in detail.

Capture the Video

When you create a DVD, you work with digital video in a DVD legal format recognized by Adobe Encore DVD. If you're working with digital video that will be supplied by external sources, make sure that the video is delivered to you in the proper format. In Chapter 1, you'll find a full breakdown of all DVD-compliant formats you can use.

If you're creating the project using video you or your client has shot using a digital camcorder, you'll have to capture the video into your PC. After capturing the digital video, you can then edit the video to compile the individual clips for your project. You can capture video using an analog capture card, but for the best results, you should use a Firewire (IEEE 1394) host controller if your camcorder has a Firewire connection. This will give you the best possible video for editing in a video-editing application. If you have the luxury of a DV deck, you can save wear and tear on the delicate mechanisms of your digital camcorder.

When you capture digital video, you may be able to specify the settings used to capture it. Adobe Premiere Pro gives you the option of using the project settings as the capture settings. If you've created a new Adobe Premiere Pro project using one of the DV-NTSC or DV-PAL presets shown next, you can capture your digital video in the same format. When capturing a project, capture the video using the television standard for your project. You should capture your audio at 48 kHz. For more information on frame sizes and audio supported by Adobe Encore DVD, refer to Chapter 2.

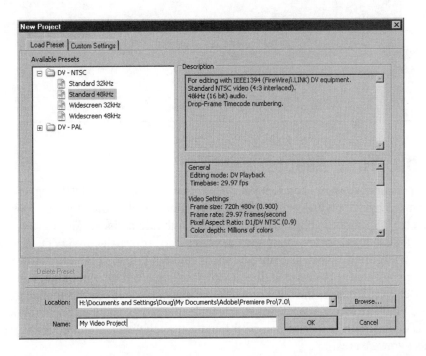

The manner in which you capture video varies, depending on the application you use. Many applications give you the option of capturing the entire DV tape, or sections of the tape. After you create a new project in Adobe Premiere Pro, choose

File | Capture, or Press F5 to open the capture dialog box. The following illustration shows video being captured with Adobe Premiere Pro:

Assemble Your Assets

If your client or coworkers present you with edited video clips in a DVD-legal format recognized by Adobe Encore DVD, you can import the clips directly into your DVD project. Make sure the clips are the proper frame rate and size for your project's television standard. After you import the clips into Adobe Encore DVD, you can perform minor edits such as trimming the in and out points of the clip. However, if you are presented with unedited digital video, or capture your own digital video for a project, you'll require the resources of a professional video editing application like Adobe Premiere Pro or Sony Vegas to assemble the footage and then render the video in a DVD-compliant format compatible with the television standard for the DVD project you are creating.

With most professional-level video-editing applications, you can work with multiple video and audio tracks. You can composite video tracks to achieve special effects, apply video transitions, apply video filters, and much more. Consult your video-editing application user manual for detailed instructions. When you create video projects in Adobe Premiere Pro, you can add markers to the timeline that you can use as Chapter Points in Adobe Encore DVD. For more information on adding markers to an Adobe Premiere Pro project, refer to Chapter 4. The following illustration shows a multitrack project being created in Adobe Premiere Pro:

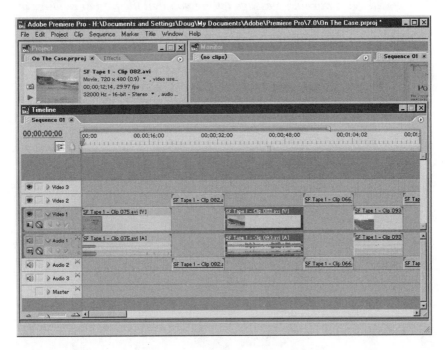

After meticulously editing your production, you're ready to render it in a format that can be recognized by Adobe Encore DVD. Depending on the application you're using, you can render one file for both audio and video, or separate files. If you're rendering an Adobe Premiere Pro project, you can render the file as an AVI file that contains both audio and video, or you can render individual files for AVI video, and Waveform (WAV) audio. You can choose the video compression codec, and specify the audio sample rate and sample type. Refer to Chapter 1 for more information on compression codecs. For the best quality sound, you should choose 48kHz, 16-bit sampling rate.

When you render video as an AVI file, the file will have to be transcoded in Adobe Encore DVD before you can build the project. If desired, you can render the file as MPEG-2 video using your DVD project's television standard. In Adobe Premiere Pro, choose File | Export | Adobe Media Encoder to open the dialog box shown next. From within this dialog box, you can choose one of the presets, or create your own custom preset. The settings are similar to the transcode settings discussed in Chapter 7. You can also choose the audio format for the exported file. You can choose to encode the audio as MPEG audio, PCM audio, or SurCode for Dolby Digital. You can encode three files with SurCode for Dolby Digital, after which time you'll be required to purchase the plug in.

NOTE

If you encode the audio as MPEG audio, this is DVD-compliant for Adobe Encore DVD. However, if you use MPEG audio when building an NTSC DVD, you will get a warning message telling you that the format requires at least one PCM or Dolby Digital audio track per timeline. You can successfully build the project with MPEG sound—however, the resulting disc may not be compatible with older set top NTSC DVD players.

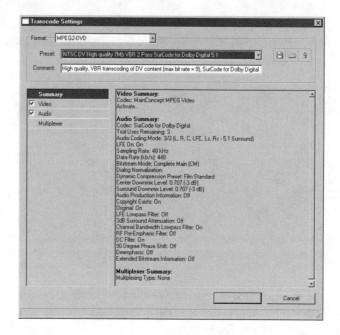

If you have the ability to render separate audio and video files, you can set up multiple audio tracks that are perfectly synchronized with the video. This option is useful if your DVD project will have alternate audio tracks in different languages, with director's comments, and so on. Render each audio track in a format recognized by Adobe Encore DVD. You can import the audio files into Adobe Encore DVD and add up to eight audio tracks per timeline, as outlined in Chapter 12. Rendering separate files is also helpful if you're going to transcode assets in Adobe Encore DVD. You have more flexibility when transcoding individual audio and video files as compared to transcoding a single video file with embedded audio.

Plan Your Project

A journey of a thousand miles begins with a single step, and the path to a compelling DVD disc begins when you decide to create a DVD project. Your path from start to finish will be different depending on the needs of your client and the complexity of your project. Therefore, you should consider planning all but the simplest DVD project before you create a new project in Adobe Encore DVD. The first and most crucial step is to make sure you and the client are on the same wavelength. Make sure you uncover all of the client's needs and expectations before beginning the project and, for that matter, before preparing a proposal. If, on the other hand, you are creating a DVD to promote yourself or your company, you should put your ideas and expectations down on paper.

After you know the basic concept of the finished project, you can begin planning the project. If you've already edited the video clips, you'll know how many timelines and chapter points will be involved. If your client has presented you with the project video, it's advisable to review the content in order to determine logical chapter points before pressing on. This information will determine how many menus will be required and how many buttons will be required on each menu. You can then sketch out the navigation for each menu as outlined in Chapter 4.

At this point in the project, you'll also have to determine the size of the media required for the project. If you've edited the footage yourself and rendered the files as MPEG video, or previewed the client's footage, you'll have a good idea of whether you'll need a single-sided or dual-sided disc for your project. If you are working with digital video that will need to be transcoded and know the desired disc size, you can calculate the required data rate by performing the calculations discussed in Chapter 4.

Create A Custom Menu

If you or your client require a state-of-the-art DVD, you can go a long way toward fulfilling that goal by creating a custom menu in Adobe Photoshop. When you create a menu in Adobe Photoshop, you are creating a document with square pixels. Begin by choosing a preset size to match your DVD project television standards. After that, use the various Adobe Photoshop tools and menu commands to create the menu. You can take advantage of Adobe Photoshop filters to add a touch of panache to the project. If you're creating a custom background comprised of bitmap images, you can color correct the images, adjust hue and saturation, and much more. You can use Adobe Photoshop's text tools to create stylized text for your DVD menus. Each text object you create is on its own layer. You can apply layer styles to text, which are preserved when you import the menu into Adobe Encore DVD.

In addition to creating sophisticated backgrounds and title text for your DVD menus, you can also create button sets in Adobe Photoshop. When you create a button in Adobe Photoshop, you create a layer set, with individual layers for the video thumbnail and pictures. Rename the layer set and inclusive layers so that Adobe Encore DVD recognizes the layer set as a button set. If you or your client requires uniquely shaped buttons, you can create them using layer or vector masks. For a detailed discussion of how to create custom menus and button sets in Adobe Photoshop, refer to Chapter 11.

A

Create A New Project

After you prepare your assets and plan your project, you're ready to create a new project. When you create a new Adobe Encore DVD project, specify project settings. Your first move is to select the television standard for your project. After doing so, Adobe Encore DVD sets the project transcode settings and you're presented with a blank canvas on which to create your DVD. If you've followed the previous steps, that canvas won't be blank for long.

When you initially create a DVD project, you can also specify other project settings such as the media size, region code, and whether the DVD disc will have copy protection or not. You specify these settings using the Disc tab, as outlined in Chapter 4. You create your DVD project using the tabs, palettes, and tools discussed in Chapter 3. But before you can put those tools to work, you must first import assets into the project.

Import Assets

After you create a new project, your next step is to import the project assets. You can import audio, video, and image assets by choosing File | Import as Asset. If you've created custom menus in Adobe Photoshop, you import these files by choosing File | Import as Menu. When you import project assets, they're neatly organized in the Project tab. If you are creating a complex project, you may find it useful to segregate the various assets of your projects by creating Project tab folders, as shown next:

A unique icon identifies each asset you import into the project. You can preview an audio, video, or image asset by selecting it in the Project tab. A thumbnail version of the asset appears in the Project tab preview window. If the asset is an audio or video asset, you can play the asset by clicking the play button to the left of the window. For more information on the Project tab and importing assets, refer to Chapter 5.

Create Timelines and Chapter Points

Before you can create navigation for your DVD project, you must first create timelines—you can create timelines for audio, video, and image assets. As a rule, you generally match an audio track with its video counterpart; however, you can use an audio-only track to segue from a video timeline to a menu, or use an audio track as an intro for a DVD.

When you create a timeline for a video or image asset, the first frame of the timeline is the chapter point. If you add additional images to an image timeline, a new chapter point is created for each image. If desired, you can delete these chapter points and play the timeline as a continuous slide show. For more information on video timelines, image timelines, chapter points, and poster frames, refer to Chapter 6.

The timeline chapter point is also the chapter point poster frame. The poster frame is displayed in the video thumbnail layer of a menu button. If the default chapter point poster frame doesn't adequately represent the content of the timeline, you can set another frame in the chapter as the poster frame.

You can set chapter points or choose frames for chapter point poster frames while viewing the timeline in the Monitor window shown on the following page. Chapter point poster frames are displayed in buttons with video thumbnail layers. If you are linking to a text-only button, you don't need to concern yourself with the poster frame.

A

When you have a lengthy timeline, you can break the timeline into chapter points, which are used as destinations for menu buttons. The following image shows a timeline with multiple chapter points. For detailed information on creating timelines, chapter points, and so on, refer to Chapter 6.

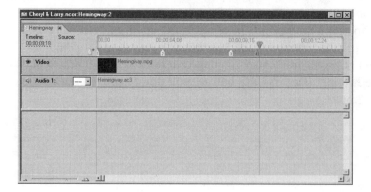

Add Alternate Audio Tracks

If desired, you can add alternate audio tracks to a timeline. When you add alternate audio tracks, you can include audio in different languages, or other tracks with features such as 5.1 Surround Sound or director's comments. You can have a total of eight audio tracks for each timeline in your project. The following image shows a timeline with alternate audio tracks:

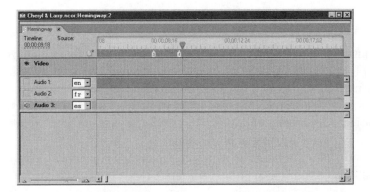

A DVD player can only play one soundtrack at a time. When you create alternate audio tracks, you need to give viewers options for choosing alternate audio tracks. Do so by specifying that a different audio track is played when a button is clicked. If you have enough alternate soundtracks, you can create a menu that viewers use to select the desired track. For more information on alternate audio tracks, refer to Chapter 12.

Add Subtitles

You can add text information to your DVD projects in the form of subtitles. Subtitles are useful when you're creating a DVD project that may be viewed by the hearing impaired. You can also add foreign language subtitles to a DVD project. You can have up to 32 subtitle tracks for each timeline.

You can import subtitles or create subtitles directly in the Monitor window. When you create your own subtitles or import a text subtitle script, you can specify the font style, font size, and other characteristics using the Character palette. You can also specify which colors are used to display the text fill and stroke, and which color is used to antialias subtitle text by choosing a timeline color set.

A set top DVD player can only display one subtitle track at a time. Therefore, when you have a DVD production with multiple subtitle tracks, you need to create menu navigation that gives viewers the option to choose which subtitle track they view. If you have enough subtitle tracks and audio tracks in a project, you can create a setup menu that gives viewers the option of which audio track they hear and which subtitle track they view. You'll find detailed information on working with subtitle tracks and creating setup menus in Chapter 13.

Create Navigation

After creating timelines for the video and image assets in your production, you create navigation for the DVD. When you create menu navigation, you create links from menu buttons to a timeline, or if the timeline has multiple chapter points to individual chapter points. Menu navigation can make or break your project. If you've got a DVD project with broken links, it's like viewing a web site with broken links. It marks the production as being unprofessional.

When you create navigation for a project, the starting point is always the project's main menu. You can use a menu from the Library as the project main menu, or a menu that you create in Adobe Photoshop. The main menu lists the major parts of your project such as a list of the scenes, setup options, and so on. Each listing in the main menu links to a submenu in the project. Each submenu contains buttons that viewers use to view individual scenes, select setup options, and so on. Menus are covered in detail in Chapter 8. Submenus and button navigation are covered in detail in Chapter 9. Creating custom menus in Adobe Photoshop is covered in detail in Chapter 11.

Use Library Menus and Buttons

The Adobe Encore DVD Library is stocked with professionally created menus and buttons. You can use these items to quickly create navigation for your DVD project. If you've created a menu in Adobe Photoshop, you can use buttons from the Adobe Encore DVD Library to create navigation for your project. You also use the Adobe Encore DVD Library to store items you've created such as custom menus, modified menus, buttons, and graphics. You'll find complete information on the Adobe Encore DVD Library in Chapter 5.

> TIP
>
> *Insert the Adobe Encore DVD CD ROM installation disc in your CD ROM drive and navigate to the Goodies folder. Here you'll find a wide array of background images, background videos, buttons, and menus you can use for your DVD projects.*

Customize and Edit Menus in Adobe Photoshop

If you add Library menus and buttons to your project, but want to customize them to suit your or your client's taste, you can do so easily by customizing the menu in Adobe Photoshop. To customize a menu, select it in the Menus tab and then choose Menu | Edit in Photoshop. After choosing this command, the selected menu is opened in Adobe Photoshop. You can then customize the menu to suit your project.

After saving the menu in Adobe Photoshop, you return to Adobe Encore DVD and all of your edits are applied to the menu. Customizing and editing menus in Adobe Photoshop are covered in Chapter 8.

Link Buttons to Video Timelines and Chapter Points

As mentioned previously, menu navigation can make or break your project. Fortunately, Adobe Encore DVD gives you many methods for creating links. You can create button links by dragging a clever device known as the pick whip from the Properties palette to a timeline. The pick whip appears as an icon to the left of several fields you'll encounter in the Properties palette. You create a link by clicking the pick whip icon and dragging it to a timeline, chapter point, or menu button. You can also create a menu link while editing a menu in the Menu Editor by dragging a timeline or chapter point from the Timelines tab to a menu button, as shown next. You'll find detailed information on using the pick whip and creating navigation links in Chapter 9.

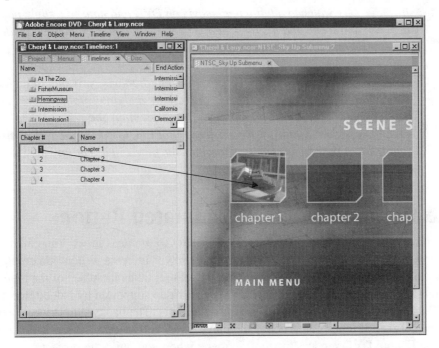

A

Set Video Timeline End Actions

When you add a timeline to a project, you create an asset that can be viewed when a user activates the button that is linked to the timeline. The timeline's default end action is stop. As a rule, you create a link for a timeline end action. This returns viewers to a menu giving them the option of viewing a different timeline or chapter point. You can set a timeline's end action from within the Properties palette by choosing a link from a drop down menu, or by dragging the Link pick whip to a button of a menu you are currently editing in the Menu Editor window.

If you have a project where you want to give viewers options, you can edit your video in a video-editing application and create a separate video clip for each important part of your presentation. In Adobe Encore DVD you create a separate timeline for each video clip. You link the end of one video clip timeline to the start of the next by specifying the next timeline as the end action link. After linking all parts of your presentation together in this manner, you create a menu option to play the entire presentation, which links to the first timeline. To give viewers an option, you create menu choices to view individual parts of the presentation. Each button links to the applicable timeline. You set each button's override action to return the viewer to a menu. In essence, the override action overrides the timeline's end action. For more information on setting end and override actions, refer to Chapter 9.

NOTE *If you've ever watched a full-length movie on DVD and noticed a slight pause between one scene and the next, that is because the movie has been divided into individual timelines for each scene. The slight pause is caused by the amount of time it takes the set top DVD player to execute the end action. If possible, when creating a DVD for a full-length movie, you should create a single timeline for the entire movie and create a chapter point where each scene begins. You can then create a button that plays the entire movie, and create individual buttons for each scene that link to the applicable chapter point. When your viewers choose to watch the movie in its entirety, they won't experience a pause between scenes.*

Create Motion Menus and Animated Buttons

You have other options to take your DVD project to the next level: You can use motion menus and animated buttons to pique viewer interest. When you create a motion menu, you create a link to a video asset that takes the place of the menu's current background. You can use any DVD-legal video supported by Adobe Encore DVD as the source for your motion menu background. The background video loops for the number of times you specify, and plays whenever a viewer selects the menu. Adobe Encore DVD renders the motion menu when you build the project.

To augment a motion menu, you can create a link to an audio file that plays whenever the menu is selected. To take the menu to the next level, you can create animated buttons. For a detailed discussion on creating motion menus, refer to Chapter 10.

Preview Your Project

After meticulously crafting your DVD project, you're ready to build the project to disc. However, before building the project to disc, you should preview your project to make sure all of the menu buttons link to the appropriate timelines, and that the proper actions occur when a timeline finishes playing. You can preview a project from a menu you are currently editing in the Menu Editor by right clicking inside the menu and then choosing Preview from Here from the shortcut menu. You can also preview the entire project from first play by choosing File | Preview Project. Whether you preview an entire project, or a single menu, it appears in the Project Preview window shown next. The Project Preview window contains buttons that simulate the buttons in a set top DVD player. You can also render the current motion menu from within the Project Preview window.

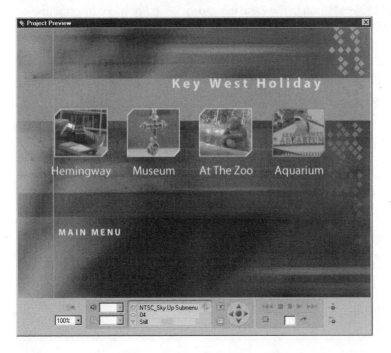

You can also test a project for broken links; links not set, orphaned menus, and orphaned timelines. You can test for any of these eventualities by choosing Edit | Check Links, or by clicking the Check Links button in the Disc tab. You'll find detailed information on the Project Preview Window and checking links in Chapter 14, along with other items you should consider before building your project.

Build Your Project

After previewing your project and determining everything is to your satisfaction, you're ready to build the project. Prior to building a project, you can change project settings by clicking the Project Settings button in the Disc tab. For example, you can add copy protection to a project, or specify a different size media. To build a project, choose File | Build DVD and then choose the desired type of build from the drop down menu. You can build a project to DVD disc using your computer's DVD burner, create a DVD folder, create a DVD image, or create a DVD Master. To create a DVD Master you must have a DLT (Digital Linear Tape) drive hooked up to your machine. You create a DVD Master when your project will be replicated by a service center. If your project has copy protection, a DVD Master is the only way you can apply the protection to duplicated or replicated discs. You can also build a project, by clicking the Build Project button in the Disc tab shown next. For more information on building your DVD projects, refer to Chapter 15.

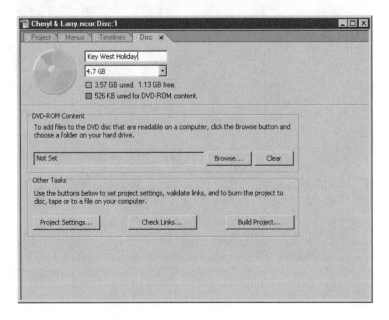

Appendix B

Adobe Encore DVD Keyboard Shortcuts

A dobe Encore DVD has a robust set of menu commands that you use when authoring a DVD project. Many of the commands have keyboard shortcuts that, once memorized, can streamline your DVD authoring workflow. Many of the keyboard shortcuts are common to all Windows programs, while others are germane to Adobe Encore DVD. The following tables list all of the Adobe Encore DVD keyboard shortcuts and are organized according to the menu on which they appear.

Command	Keyboard Shortcut
New Project	CTRL-N
Open Project	CTRL-O
Close	CTRL-W
Close Project	SHIFT-CTRL-W
Save	CTRL-S
Save As	SHIFT-CTRL-S
New Folder	SHIFT-CTRL-N
Import As Asset	CTRL-I
Import As Menu	SHIFT-CTRL-I
Replace Asset	CTRL-H
Locate Asset	SHIFT-CTRL-H
Preview	ALT-CTRL-SPACE
Exit	CTRL-Q

TABLE B-1   File Menu Keyboard Shortcuts

Command	Keyboard Shortcut
Undo	CTRL-Z
Redo	SHIFT-CTRL-Z
Cut	CTRL-X
Copy	CTRL-C
Paste	CTRL-V
Duplicate	CTRL-D
Paste as Subpicture	SHIFT-CTRL-V
Rename	SHIFT-CTRL-R
Clear	DELETE
Edit Original	CTRL-E
Select All	CTRL-A
Deselect All	SHIFT-CTRL-A
Check Links	SHIFT-CTRL-L
Preferences I General	CTRL-K

TABLE B-2   Edit Menu Keyboard Shortcuts

Command	Keyboard Shortcut
Convert To Button	CTRL-B
Convert To Object	SHIFT-CTRL-B
Create Subpicture	ALT-CTRL-B
Link To	CTRL-L
Drop Shadow	SHIFT-CTRL-O
Arrange Menu Group	
Bring To Front	SHIFT-CTRL-]
Bring Forward	CTRL-]
Send Backward	CTRL-[
Send To Back	SHIFT-CTRL-[

TABLE B-3   Object Menu Keyboard Shortcuts

Command	Keyboard Shortcut
New Menu	CTRL-M
Edit Menu In Photoshop	SHIFT-CTRL-M

TABLE B-4   Menu Menu Keyboard Shortcuts

Command	Keyboard Shortcut
New Timeline	CTRL-T
Bring Tracks Into View	ALT-CTRL-T
Add Chapter Point	CTRL-F1
Set Poster Frame	SHIFT-CTRL-F1
Save Frame As File	SHIFT-CTRL-F
Play Timeline	CTRL-SPACE or SPACE

TABLE B-5   Timeline Menu Keyboard Shortcuts

B

Command	Keyboard Shortcut
Zoom In	CTRL-+
Zoom Out	CTRL--
Fit In Window	CTRL-0
Actual Size	CTRL-1
Show Menu Image	CTRL-2
Show Normal Subpicture	CTRL-3
Show Selected Subpicture	CTRL-4
Show Activated Subpicture	CTRL-5
Show Safe Areas	CTRL-6
Show Button Routing	CTRL-7

TABLE B-6   View Menu Keyboard Shortcuts

Menu Command	Keyboard Shortcut
Properties	F5
Character	F6
Layers	F7
Library	F8
Project	F9
Menus	F10
Timeline	F11
Disc	F12
Monitor	SHIFT-CTRL-SPACE

TABLE B-7   Window Menu Keyboard Shortcuts

Menu Command	Keyboard Shortcut
Adobe Encore Help	F1

TABLE B-8   Help Menu Keyboard Shortcut

Appendix C

Internet DVD and Video Resources

You can find out just about anything on any topic by surfing the Internet. The Internet is indeed a treasure trove of information. However, if you type a single keyword in your favorite search engine, you'll be overwhelmed with a plethora of URLs that may or may not contain the information you require. You have better things to do with your time (like author DVDs for fun and profit) than sift through a seemingly endless list of web sites. In this section you'll find some bona fide resources for information about DVDs and Digital video. The URLs for these sites have been verified at the time of this writing. However, the Internet is in a constant state of flux and what is here today may be gone tomorrow.

Adobe Resources

Adobe's web site is a vast resource for the digital videographer, digital photographer, web designer, and graphic designer. You'll find information for all of the Adobe products mentioned in this book as well as support resources, tips and tutorials, and other information. The home page of Adobe's web site is located at www.adobe.com. Listed next are some relevant resources for Adobe Encore DVD and digital photographers.

Adobe Encore DVD Support Page (www.adobe.com/support/products/encore.html)

In this section of Adobe's web site you'll find the latest information about Adobe Encore DVD. You'll find links to tutorials, Adobe Encore DVD support, and a link to a section where you can request a feature for future releases of Adobe Encore DVD. You'll also find a section devoted to Top Issues. If you have any questions about Adobe Encore DVD, this is the right place to find the answer.

Adobe Studio Tips and Tutorials (http://studio.adobe.com/tips/main.jsp?DEPLOY)

This URL is a portal to tips and tutorials for Adobe's entire product line. The first time you access this resource, you'll be required to register, but don't worry—it's free and registering won't fill your inbox with Spam. After registering, you'll have access to a wide variety of expert tutorials for Adobe's entire product line.

> TIP   *If you own Adobe Acrobat Standard or Adobe Acrobat Professional, you can capture a web page as a PDF from within the Internet Explorer. This is a wonderful way to save a large web tutorial for future reference. Just make sure you're not violating any copyright laws by capturing the page, and don't distribute the information to others.*

Adobe Digital Video Page (www.adobe.com/motion/main.html)

This web page is the home page for Adobe's digital video products. If you've considered expanding your software resources with additional video applications, you'll find information about all of Adobe's digital video solutions here. In addition to finding details about digital video products and other useful information, you'll also find links to tryout versions of these powerful applications.

Adobe Digital Imaging (www.adobe.com/digitalimag/main.html)

In this section of Adobe's web site you'll find information for every Adobe digital-imaging application. You'll also find a download link that opens a page where you can download updates, plug-ins, and filters for your Adobe products. If you've contemplated purchasing one of Adobe's digital-imaging applications, you'll find a tryout link on the Downloads page. If you have Adobe Photoshop as part of your graphics software suite, you'll be especially interested in the Actions you can download and add to your copy of Adobe Photoshop.

DVD Resources

DVD-authoring software and hardware has been increasing in popularity. Whenever a topic becomes hot, you can expect to find web sites popping up quicker than gubernatorial candidates in a California recall election. The following are just a few of the useful DVD resources you can find on the Internet.

Creativecow.com (www.creativecow.com)

At this site you'll find an Adobe Encore DVD forum. In addition, you'll find a wealth of information on many other Adobe products, including Adobe Premiere Pro, Adobe After Effects, and more. You'll find sections for tutorials and articles about all of these applications and more.

DVDRhelp.com (www.dvdrhelp.com)

At this web site you'll find resources and information about authoring DVDs. You'll find guides and articles about DVD authoring, as well as information on digital video. You'll also find a links section, filled to overflowing with links to other web resources on DVDs and digital video.

DVD Creation (www.dvdcreation.com)

This web site offers information about DVD-related products and applications, as well as news, tutorials, and other resources. You'll find links to forums, including forums for popular Adobe applications, and, of course, Adobe Encore DVD.

C

Videomaker.com (www.videomaker.com)

Here's yet another useful resource with information and tips about digital video. You'll also find information about DVD hardware and other equipment related to digital video and a section called DVD Central that contains information about DVD tools, techniques, production news, and a DVD forum.

VideoGuys.com (www.videoguys.com)

In this site you'll find the best of both worlds: information about DVD authoring and digital video products. You can purchase DVD and video equipment at this web site as well as software applications. In the DVD Cookbook section, you'll find online resources for DVD authoring.

Art Beats.com (www.artbeats.com)

If you're looking for a source of royalty-free video clips you can use as backgrounds for your motion menus, you'll find lots of choices here. You can purchase royalty-free video collections of just about any imaginable subject in both NTSC and PAL format.

Digital Video Resources

If you create digital video for your DVD projects, you can find a wealth of information on the Internet. There is an abundance of web sites with information on this popular subject. The following sites are an attempt to separate the wheat from the chaff, but this collection of sites is by no means an authoritative resource of web sites pertaining to digital video.

Digital Video Professionals Association (www.dvpa.com)

At this web site you'll find a wealth of information concerning digital video. After registering with the site, you'll have access to tutorials, a job bank, and a gallery of other members' works.

Filmmaker Magazine (www.filmmakermagazine.com)

This is the online portal for Filmmaker Magazine. If you create DVD projects for independent filmmakers, you may find information of interest at this site. You'll also find an extensive list of resources for independent filmmakers.

Shockwave Sound.com (www.shockwave-sound.com)

If you need a resource for royalty-free sound loops for your video productions, or for audio tracks for your DVD projects, you'll find them here. You can purchase and download royalty-free loops and sound samples for your productions. You'll also find a Resources section at this site that features links to sound tutorials and other useful information.

Sounddogs.com (www.sounddogs.com)

Here's another web resource chock full of sound effects, loops, samples, and vocals you can use in video productions. You can also use these sounds with an application like Adobe Audition to augment a soundtrack for a DVD project with loops and sound effects.

Cyberfilmschool.com (www.cyberfilmschool.com)

Here's a web site packed with useful information for digital filmmakers. You'll find articles, tips, and techniques, plus information on creating DVDs.

Wrigley Video Productions (www.wrigleyvideo.com)

At this web site you'll find an Adobe Premiere tutorial section with an impressive collection of tutorials. If you'd like to swap ideas and information with other Adobe Premiere users, check out their user forum.

Digitalfilmmaker.net (www.digitalfilmmaker.net)

This site is a consortium of independent filmmakers, photojournalists, and videojournalists. You can find some outstanding information on the art of digital video here.

C

Index

References to figures and illustrations are in italics.

INTERNATIONAL CONTACT INFORMATION

AUSTRALIA
McGraw-Hill Book Company
Australia Pty. Ltd.
TEL +61-2-9900-1800
FAX +61-2-9878-8881
http://www.mcgraw-hill.com.au
books-it_sydney@mcgraw-hill.com

CANADA
McGraw-Hill Ryerson Ltd.
TEL +905-430-5000
FAX +905-430-5020
http://www.mcgraw-hill.ca

GREECE, MIDDLE EAST, & AFRICA
(Excluding South Africa)
McGraw-Hill Hellas
TEL +30-210-6560-990
TEL +30-210-6560-993
TEL +30-210-6560-994
FAX +30-210-6545-525

MEXICO (Also serving Latin America)
McGraw-Hill Interamericana Editores
S.A. de C.V.
TEL +525-1500-5108
FAX +525-117-1589
http://www.mcgraw-hill.com.mx
carlos_ruiz@mcgraw-hill.com

SINGAPORE (Serving Asia)
McGraw-Hill Book Company
TEL +65-6863-1580
FAX +65-6862-3354
http://www.mcgraw-hill.com.sg
mghasia@mcgraw-hill.com

SOUTH AFRICA
McGraw-Hill South Africa
TEL +27-11-622-7512
FAX +27-11-622-9045
robyn_swanepoel@mcgraw-hill.com

SPAIN
McGraw-Hill/
Interamericana de España, S.A.U.
TEL +34-91-180-3000
FAX +34-91-372-8513
http://www.mcgraw-hill.es
professional@mcgraw-hill.es

UNITED KINGDOM, NORTHERN, EASTERN, & CENTRAL EUROPE
McGraw-Hill Education Europe
TEL +44-1-628-502500
FAX +44-1-628-770224
http://www.mcgraw-hill.co.uk
emea_queries@mcgraw-hill.com

ALL OTHER INQUIRIES Contact:
McGraw-Hill/Osborne
TEL +1-510-420-7700
FAX +1-510-420-7703
http://www.osborne.com
omg_international@mcgraw-hill.com

Know How

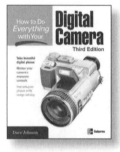

How to Do Everything with Your Digital Camera
Third Edition
ISBN: 0-07-223081-9

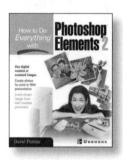

How to Do Everything with Photoshop Elements 2
ISBN: 0-07-222638-2

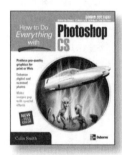

How to Do Everything with Photoshop CS
ISBN: 0-07-223143-2
4-color

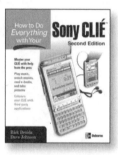

How to Do Everything with Your Sony CLIÉ
Second Edition
ISBN: 0-07-223074-6

How to Do Everything with Macromedia Contribute
0-07-222892-X

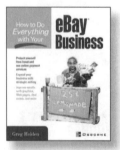

How to Do Everything with Your eBay Business
0-07-222948-9

How to Do Everything with Illustrator CS
ISBN: 0-07-223092-4
4-color

How to Do Everything with Your iPod
ISBN: 0-07-222700-1

How to Do Everything with Your iMac,
Third Edition
ISBN: 0-07-213172-1

How to Do Everything with Your iPAQ Pocket PC
Second Edition
ISBN: 0-07-222950-0